148
Structure and Bonding

AF768078

Series Editor:

D.M.P. Mingos, Oxford, United Kingdom

Editorial Board:

F.A. Armstrong, Oxford, United Kingdom
X. Duan, Beijing, China
L.H. Gade, Heidelberg, Germany
K.R. Poeppelmeier, Evanston, IL, USA
G. Parkin, NewYork, USA
M. Takano, Kyoto, Japan

For further volumes:
http://www.springer.com/series/430

Aims and Scope

The series *Structure and Bonding* publishes critical reviews on topics of research concerned with chemical structure and bonding. The scope of the series spans the entire Periodic Table and addresses structure and bonding issues associated with all of the elements. It also focuses attention on new and developing areas of modern structural and theoretical chemistry such as nanostructures, molecular electronics, designed molecular solids, surfaces, metal clusters and supramolecular structures. Physical and spectroscopic techniques used to determine, examine and model structures fall within the purview of *Structure and Bonding* to the extent that the focus is on the scientific results obtained and not on specialist information concerning the techniques themselves. Issues associated with the development of bonding models and generalizations that illuminate the reactivity pathways and rates of chemical processes are also relevant.

The individual volumes in the series are thematic. The goal of each volume is to give the reader, whether at a university or in industry, a comprehensive overview of an area where new insights are emerging that are of interest to a larger scientific audience. Thus each review within the volume critically surveys one aspect of that topic and places it within the context of the volume as a whole. The most significant developments of the last 5 to 10 years should be presented using selected examples to illustrate the principles discussed. A description of the physical basis of the experimental techniques that have been used to provide the primary data may also be appropriate, if it has not been covered in detail elsewhere. The coverage need not be exhaustive in data, but should rather be conceptual, concentrating on the new principles being developed that will allow the reader, who is not a specialist in the area covered, to understand the data presented. Discussion of possible future research directions in the area is welcomed.

Review articles for the individual volumes are invited by the volume editors.

In references *Structure and Bonding* is abbreviated *Struct Bond* and is cited as a journal.

Jan C.A. Boeyens • Peter Comba
Editors

Electronic Structure and Number Theory

Bohr's Boldest Dream

With contributions by

Jan C.A. Boeyens · Peter Comba · Demetrius C. Levendis ·
C.J.H. Schutte

Springer

Editors
Jan C.A. Boeyens
Unit for Advanced Scholarship
University of Pretoria
Pretoria
South Africa

Peter Comba
Anorganisch-Chemisches Institut
Universität Heidelberg
Heidelberg
Germany

ISSN 0081-5993 ISSN 1616-8550 (electronic)
ISBN 978-3-642-43519-5 ISBN 978-3-642-31977-8 (eBook)
DOI 10.1007/978-3-642-31977-8
Springer Heidelberg New York Dordrecht London

© Springer-Verlag Berlin Heidelberg 2013
Softcover re-print of the Hardcover 1st edition 2013
This work is subject to copyright. All rights are reserved by the Publisher, whether the whole or part of the material is concerned, specifically the rights of translation, reprinting, reuse of illustrations, recitation, broadcasting, reproduction on microfilms or in any other physical way, and transmission or information storage and retrieval, electronic adaptation, computer software, or by similar or dissimilar methodology now known or hereafter developed. Exempted from this legal reservation are brief excerpts in connection with reviews or scholarly analysis or material supplied specifically for the purpose of being entered and executed on a computer system, for exclusive use by the purchaser of the work. Duplication of this publication or parts thereof is permitted only under the provisions of the Copyright Law of the Publisher's location, in its current version, and permission for use must always be obtained from Springer. Permissions for use may be obtained through RightsLink at the Copyright Clearance Center. Violations are liable to prosecution under the respective Copyright Law.
The use of general descriptive names, registered names, trademarks, service marks, etc. in this publication does not imply, even in the absence of a specific statement, that such names are exempt from the relevant protective laws and regulations and therefore free for general use.
While the advice and information in this book are believed to be true and accurate at the date of publication, neither the authors nor the editors nor the publisher can accept any legal responsibility for any errors or omissions that may be made. The publisher makes no warranty, express or implied, with respect to the material contained herein.

Printed on acid-free paper

Springer is part of Springer Science+Business Media (www.springer.com)

Preface

At the time of writing, the world of physics is buzzing with the identification of a Higgs field. In the popular press, which refers to the Higgs boson as the "God" particle, we read [1] that

> It is the last unobserved piece of the Standard Model, the most convincing explanation available for the way the universe works... The purpose of the Higgs boson is to inculcate mass into those particles which weigh something... The search for the Higgs is a search for closure on the old world. Supersymmetry is the new. It might also (explain) "dark matter".

These are brave words, which echo the expectation of another theorist [2]:

> For some physicists, the end of the road is in sight, and the main question is simply, How long until we reach it?

In view of the common contention that all of chemistry is reducible to physics it may therefore seem expedient to explore the implications of this mighty discovery on our understanding of molecules and their structure. However, the immediate simple answer is zero.

The standard model of physics is based entirely on dimensionless point particles, and, whatever it may reveal about dark matter, it offers no explanation of the extension and structure of molecules. These elementary particles acquire their mass mathematically, on interaction with the hypothetical Higgs field, in a process of [2]

> miraculous mass generation.

The nature and source of mass remain unexplored. Extrapolation from the Higgs field to molecular shape is just a bridge too far.[1]

The mathematics of the standard model relies on spontaneous breaking of a gauge symmetry, defined by the phase of an elementary-particle wave function. Theoretical chemists, guided by molecular physicists, have imprudently adopted the same model, without a thorough physical basis, primarily for computational

[1] The Higgs is an incomprehensible abstraction, a partial solution to an extraordinarily rarified and perhaps always incomplete intellectual puzzle - Daniel Sarewitz, Nature 488:431 (23 August 2012).

purposes. In physics it serves for the computation of observable spectroscopic and fragmentation patterns, but in the chemical context robust molecular properties can only be simulated *empirically* as probabilistic particle distributions. As a theory of molecular structure and chemical interaction this pursuit is not very useful.

In this volume an alternative to quantum chemistry is proposed. At the outset we challenge the idea of gauge invariance associated with structureless point particles as a logical impossibility. All known phase relationships are associated with wave systems and all waves are distinguished by periodic variation specified by integers. These are precisely the properties featured as textbook characteristics of quantum systems.

The sordid infighting among quantum theorists to gain ascendency of the particle or Copenhagen model over the wave alternative may be irrelevant. What was claimed as the final formulation of quantum physics has long since been superseded by Higgs fields and string theory. There is no longer any danger in reviving a wave model of matter in situations where it works. However, with hindsight it is now more appropriate not to reconsider a simplistic three-dimensional model, but to rather take note of the topology of space–time and the theory of general relativity at the same time.

A special theory of relativity was developed in the first place to account for the prediction of Maxwell's equations that light rays are observed to propagate at constant speed in the vacuum, irrespective of the relative motion of an observer. The resulting kinematic theory, known as Lorentz transformation, is readily shown to amount to complex rotation in four-dimensional space–time. Extension to accelerated frames requires reformulation of the theory in non-Euclidean four-dimensional space–time. Although the new emerging concepts of mass–energy equivalence, time dilation and the gravitational field have captured the imagination of the world, another important result, the equivalence of space and time variables is rarely fully appreciated. This is the seminal property that allows formulation of a wave theory of matter in four dimensions. It is also one aspect of relativity theory which is often ignored. It is difficult to properly visualize non-Euclidean four-dimensionally curved space–time, which is called for in order to take the next step forward in physics.

Some of the seven billion inhabitants of the planet still entertain the atavistic notion of an infinite flat earth, some accept a two-dimensional planetary surface, closed in three-dimensional space. However, despite the evidence from general relativity it is safe to say that, without much exception, everybody still considers cosmic space as infinite and flat in three dimensions.

As for particle physics there is also a standard model for cosmology which interprets the observed spectroscopic red shifts of galactic light as a Doppler effect caused by the expansion of three-dimensional Euclidean space, claimed to be consistent with general relativity. In actual fact general relativity demands not only the entanglement of space and time coordinates in a four-dimensional continuum but also non-Euclidean cosmic topology. The property of curvature is embodied directly in Einstein's relativistic field equations which balance the mass–energy content of the universe against its curvature. It is immediately obvious that flat space contains no matter, and given the observable matter content of the universe, space–time has to be curved and in all likelihood topologically closed. This means that the standard models of physics and cosmology are dealing with artefacts created by the use of a dimensionally and topologically fallacious model.

Topologically closed space–time has the philosophical advantage of avoiding many bothersome infinities. The universe becomes of finite extent and eternity, like a circle, has no beginning or end. Neither the finite big-bang age of the universe nor John Locke's beginning in the infinite past are attractive propositions.

When contemplating the formulation of four-dimensional theories the first measure would be the use of Minkowski space–time, which is tangent to the underlying curved manifold and adequate, to first approximation, for the analysis of macroscopic local phenomena. At the sub-atomic or galactic level the effects of curvature cannot be ignored.

With a wave model in mind as a chemical theory it is helpful to first examine wave motion in fewer dimensions. In all cases periodic motion is associated with *harmonic* functions, best known of which are defined by Laplace's equation in three dimensions. It occurs embedded in Schrödinger's equation of wave mechanics, where it generates the complex surface-harmonic operators which produce the orbital angular momentum eigenvectors of the hydrogen electron. If the harmonic solutions of the four-dimensional analogue of Laplace's equation are to be valid in the Minkowski space–time of special relativity, they need to be Lorentz invariant. This means that they should not be separable in the normal sense of Sturm–Liouville problems. In standard wave mechanics this is exactly the way in which space and time variables are separated to produce a three-dimensional wave equation.

In the same way that two-dimensional harmonics are complex functions, four-dimensional harmonics are hypercomplex functions or *quaternions*, also known as spin functions. A spin function represents the four-dimensional analogue of the conserved quantity known as angular momentum in three dimensions. The problem with standard wave mechanics is that on separation of the variables to create a three-dimensional Sturm–Liouville system the spin function breaks down into orbital angular momentum and one-dimensional spin, which disappears in the three-dimensional formulation.

It is significant to note that the spin function also describes the complex rotation which defines the Lorentz transformation. An unexpected bonus of a four-dimensional quantum mechanics is this natural merger with relativity theory, the importance of which is beyond the scope of this work. Of more immediate relevance are the relativistic ramifications in chemical systems.

An almost forgotten issue is the proposed relativistic nature of an electron as elucidated by Lorentz. The electron was seen as a flexible spherical unit of charge which distorts as it contracts in the direction of any motion. To account for the relativistic contraction of macroscopic bodies Lorentz further assumed that the electrical forces which bind atoms together were essentially states of stress and strain in the aether. Countless prominent scientists have expressed similar views without trying to develop a coherent theory of matter. The Lorentz electron model antedates de Broglie's postulate of matter waves and the development of general relativity. Reinterpreted against this background the Lorentz electron emerges as a relativistic invariant wave structure which is generated as a persistent elementary distortion of the aether, due to space–time curvature, and is described mathematically by the four-dimensional spin function.

We are reminded by Bohm [3]:

> . . . that scientific investigation is basically a mode of extending our *perception* of the world, and not mainly a mode of obtaining *knowledge* about it.

The world as perceived in four-dimensional space–time is fundamentally different from the perception in tangent space. Even the most fundamental perceptions such as the Rydberg–Ritz formula are perceived differently in four dimensions, as demonstrated by Casper Schutte in this volume. Bohm [3] identifies the problem with new ideas as

> . . . the difficulty of enertaining new concepts which clash with older ones that we have held habitually since childhood. . .

Whoever reads this volume without rejecting the picture of a point electron that only shows up as a probability distribution has the same problem. In our perception it occurs, like other elementary entities, as a persistent, flexible, wave-like, chiral distortion of space–time. It has mass, charge and spin by virtue of a characteristic wave structure. It disperses into the vacuum on interaction with another of opposite chirality.

It is no accident that both wave motion and the fundamental theory of chemistry are best described in terms of natural numbers. However, conventional wave mechanics in three dimensions offers only a partial elucidation of the periodic table of the elements. On the other hand, a detailed reconstruction, also of the more general periodicity of stable nuclides, derives directly from elementary number theory. It shows, in addition, how the periodic function responds to the state of space–time curvature and identifies the golden ratio as a possible parameter that links perceptions in tangent space to the situation in curved space–time.

This observation is exploited in the analysis of chemical systems without formal solution of the four-dimensional problem. Most of the work in this volume reports results of this type. To be of practical use the numbertheory results, which pertain specifically to pairwise diatomic interactions, might be incorporated into algorithms that simulate more complex molecules. Molecular mechanics appears to be an appropriate model to achieve this. In the current quantum chemistry world, molecular mechanics appears to be a "low-level", entirely empirical model which fully ignores the relevant electronic effects [4]. However, via specific atom types, the electronics may be accounted for, and there are a number of recent developments which include specific electronic terms, based on classical quantum chemistry [5–8]. More importantly, in terms of structural modeling, the quality of MM only depends on the force field and its parameterization based on experimental data [9]. The approach presented in this volume suggests that number-theory-based models may significantly reduce the parameter space and, more importantly, produce generic parameters, which might or might not be optimized, based on experimental data in order to obtain a higher accuracy.

The development of chemical theory in terms of number theory is not new [10]. What is new is the realization that we are dealing with a four-dimensional problem, the analysis of which requires a fundamentally different mathematical treatment. Physically it means that a 3D analysis in terms of point particles is necessarily incomplete. The analysis by Schutte demonstrates beyond any doubt that the basic assumption of atomic spectroscopy, based on the traditional separation of space and time variables, is not supported by experiment. The most glaring demonstration that a 4D analysis is required is provided by the appearance of electron spin, which never emerges in any 3D model. It is important to understand that spin is not a relativistic effect, but derives from the four-dimensional conservation of angular momentum. These features were not known before and this is the first effort to take their consequences into account. It needs a radically new model of atomic electron distribution, and the exciting thing is that such a model is provided by an optimization by logarithmic spirals. The result is in striking agreement with a standing-wave model of electron density that involves the golden ratio and many other aspects of elementary number theory. Without further assumptions this model of atomic structure can be interpreted directly as the definitive basis of atomic ionization radii and electronegativity. Our reference to previous conclusions about the origin of the periodic table is done to explain the earlier empirical conclusions more logically.

What is new in the re-examination of covalent interactions is that the approach in terms of a four-dimensional wave structure leads to a precise definition of bond order, not achieved before. Together with the new freeatom ionization radii the parameters of interatomic distance, dissociation energy, stretching force constants and diatomic dipole moments can now be derived as simple functions of the ionization radii and the golden ratio. These results have nothing in common with the more approximate simulations described before.

The papers in this volume address a single theme and in order for each of them to constitute a self-contained unit a fair amount of repetition is inevitable, even with generous cross referencing. Many readers, less familiar with the concepts of number theory, the golden mean, logarithmic spirals, hypercomplex numbers, projective geometry, general relativity and quantum field theory, may actually find some reiteration to be of benefit. Not to alienate such readers at the outset the editor in chief, Mike Mingos, patiently guided the composition of the opening paper into the style of *Structure and Bonding*. His much appreciated advice is gratefully acknowledged.

We expect our proposed approach to the understanding of chemically important issues to be rejected by many readers who operate in a comfort zone defined by probability densities, Born–Oppenheimer systems, hybrid orbitals, potential-energy surfaces, ab initio theory and DFT simulations—all of them Copenhagen spin-offs. We realize, of course, that these models have been developed to standards, where they produce a very accurate optimization of structures and properties of molecular compounds and materials in many areas [11], and for application-oriented theoreticians as well as for experimentalists, who use applied theory for predictions and interpretations, there is no immediate need to leave the comfort zone. At the same time it is interesting to note that we are not alone in advocating a rethink of theoretical chemistry. An increasing number of papers in the recent literature (e.g. [12]) call for an update of the teaching of valence theory, for the orbital concept and related ideas to be abandoned and specifically to be removed from undergraduate curricula. We do not fully agree with all those suggestions, but some ideas pioneered in this volume could hopefully find their way into a new paradigm.

Acknowledgements We acknowledge the staunch support of Martha and Maria.

References

1. The Economist (2011) December 17th:139
2. Fritsch H (1984) Quarks. Penguin Books, London
3. Bohm D (1996) The special theory of relativity. Routledge, New York
4. Bersuker IB (1996) Electronic structure and properties of transition metal compounds. Wiley, New York
5. Comba P, Zimmer M (1994) Inorg Chem 33:5368
6. Comba P, Hambley TW, Ströhle M (1995) Helv Chim Acta 78:2042
7. Root DM, Landis CR, Cleveland T (1993) J Am Chem Soc 115:4201
8. Deeth RJ (2004) Struct Bonding 113:37
9. Comba P, Kerscher M (2009) Coord Chem Rev 253:564
10. Boeyens JCA (2008) Chemistry from first principles. Springer, Dordrecht, Heidelberg, London, New York
11. Comba P (ed) (2011) Modeling of molecular properties. Wiley-VCH, Weinheim, Germany
12. Pritchard HO (2012) J Chem Ed 89:301

Pretoria, South Africa — Jan C.A. Boeyens
Heidelberg, Germany — Peter Comba

Contents

Struct Bond (2013) 148: 1–24
DOI: 10.1007/978-3-642-31977-8_1
© Springer-Verlag Berlin Heidelberg 2013

Chemistry by Number Theory

Jan C.A. Boeyens and Peter Comba

Abstract Aspects of elementary number theory pertaining to the golden ratio and the golden spiral are shown to be related to and therefore of importance in the simulation of chemical phenomena. Readily derived concepts include atomic structure, electronegativity, bond order, the theory of covalent interaction and aspects of molecular chirality. The physical interpretation of the results implicates the 4D structure of space–time as a fundamental consideration. The implied classical nature of 3D molecular structure identifies molecular mechanics as an ideal method for structure optimization, based on parameters obtained by number theory. All results point at a 4D wave structure of electrostatic charge.

Contents

J.C.A. Boeyens (✉)
Unit for Advanced Scholarship, University of Pretoria, Lynnwood Road, 0002 Pretoria, South Africa
e-mail: jan.boeyens@up.ac.za

P. Comba
Anorganisch-Chemisches Institut, Universität Heidelberg, Im Neuenheimer Feld 270, 69120 Heidelberg, Germany
e-mail: peter.Comba@aci.uni-heidelberg.de

1 Introduction

One of the most mysterious observations in Nature is the appearance of a single parameter that determines the macroscopic structure of a large variety of apparently unrelated objects, such as the distribution of florets in a composite seed head [1], the periodic table of the elements [2], the flight path of a predator bird in pursuit of its prey, the curvature of a kudu horn [1] and the surface features of a nanoparticle [3]. This ubiquitous parameter, known as the *golden ratio*, has also been called the *divine proportion* and for millennia has been used in architectural design, as a measure of human anatomical features, in works of art and in musical composition [1].

The periodic table of the elements is the single most important concept in chemistry, and we consider its correlation with the golden ratio as compelling evidence that number theory could provide a significant basis for theoretical chemistry. In order to explore this contingency, it is necessary to examine the relevant properties of the remarkable golden parameter in more detail.

2 The Golden Ratio

Mathematically, the golden ratio is formulated by the division of a unit line into two segments. An intermediate point on the line [ABC] divides the line in mean and extreme ratios AB/BC and BC/AB, $BC > AB$.

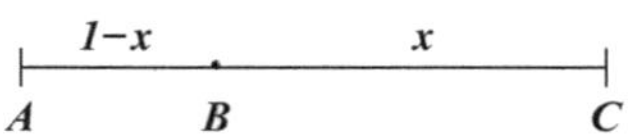

At the point where

$$\frac{AB}{BC} = \frac{BC}{AC}; \qquad \frac{1-x}{x} = \frac{x}{1},$$

the section is described by the quadratic equation

$$x^2 = 1 - x \tag{1}$$

i.e.

$$2x = -1 \pm \sqrt{5}$$
$$x = 0.61803\ldots \text{ or } -1.61803\ldots$$

This so-called golden section or golden ratio is defined by either of the irrational numbers

$$\tau = 0.61803\ldots \quad \text{or} \quad \Phi = 1.61803\cdots = \frac{1}{\tau}$$

This result follows on substituting $x = 1/x$ in (1). Multiplication by τ^n,

$$\tau^{n+2} = \tau^n - \tau^{n+1} ,$$

shows that any power of τ (or Φ) can be written as the difference between smaller powers. For example,

$$\begin{aligned}
\tau^6 &= \tau^4 - \tau^5 \\
&= 2\tau^4 - \tau^3 \\
&= 2(\tau^2 - \tau^3) - (\tau - \tau^2) \\
&= 2(2\tau^2 - \tau) - (2\tau - 1) \\
&= 2(2 - 3\tau) - (2\tau - 1) \\
&= 5 - 8\tau
\end{aligned}$$

Any power reduces to this form, in which the coefficients are successive terms in the Fibonacci series

$$0,\ 1,\ 1,\ 2,\ 3,\ 5,\ 8,\ 13,\ 21, \ldots$$

in which any term is the sum of the two preceding terms, i.e.,

$$F_{n+1} = F_n + F_{n-1}, \quad n > 0,\ F_1 = 1 .$$

A general power of τ therefore becomes

$$\tau^n = F_{n-1} - F_n\tau.$$

It follows that

$$\lim_{n\to\infty} \frac{F_{n-1}}{F_n} = \lim_{n\to\infty} \left(\frac{\tau^n}{F_n} + \tau \right) = \tau.$$

The golden ratio is expressed as a trigonometric function in the form

$$\tau = 2\cos\left(\frac{2\pi}{5}\right), \qquad \Phi = 2\cos\left(\frac{\pi}{5}\right).$$

It is therefore not surprising that the golden mean turns up in problems of fivefold symmetry. In particular, it is found that the diagonal of a unit pentagon

$$d = \sqrt{1 + 1 - 2\cos(3\pi/5)} = \Phi$$

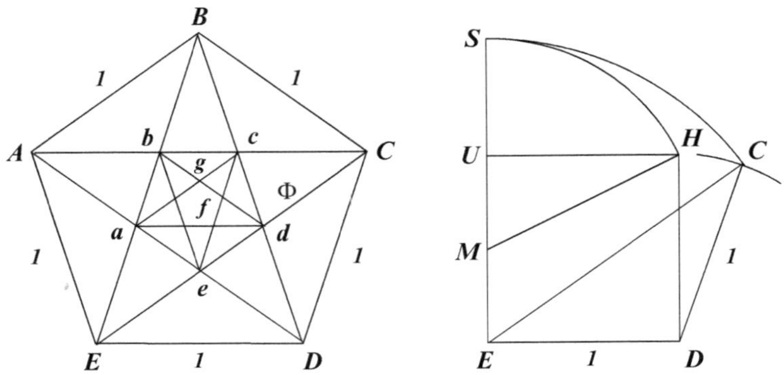

Fig. 1 Diagram to demonstrate Euclid's construction of a regular pentagon and the golden section

has the measure of the golden mean, as in Fig. 1. By noting that the quadrilateral $ABCe$ is a parallelogram and that $\triangle ACe$ and $\triangle DEe$ are similar triangles, it follows that

$$\frac{AD}{Ae} = \frac{Ae}{eD},$$

which means that the intersecting diagonals divide one another in golden ratio.

Euclid used this property to construct a regular pentagon, starting from a unit square. A circle, centered at the midpoint (M) of one side, intersects the extension of that side at S. By construction,

$$MS = MH = \sqrt{5}MU = SU + \frac{1}{2} = \Phi - \frac{1}{2}.$$

Hence, $SU = \Phi - 1 = \tau$ and $AC = \Phi$, the diagonal of a unit pentagon. This process, repeated twice, yields the complete pentagon.

The distances BD, Bd, Bc, cd, cf and fg are in geometrical progression and equal to

$$\Phi,\ 1,\ \Phi^{-1}(\tau),\ \Phi^{-2}(\tau^2),\ \tau^3,\ \tau^4,\ \text{etc.}$$

The construction of smaller or larger pentagon around the central pentagon can be continued indefinitely to define an infinite geometrical series based on τ:

$$S = \{\Phi^n,\ n = -\infty, \infty\} \tag{2}$$

and the corresponding infinite structure consisting of self-similar pentagon.

In the same way, addition of a line segment of length τ to a unit line yields the extended line [ABC], again divided in golden ratio by the intermediate point B, as before. This process, when continued indefinitely, generates a sequence of larger and larger copies of the original line in golden section. The same process unfolds in

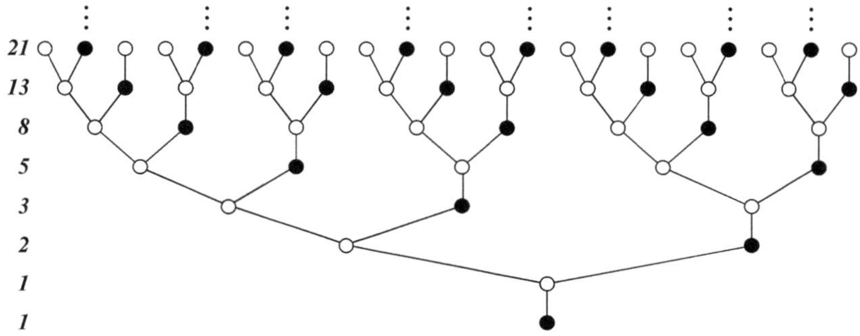

Fig. 2 The Fibonacci tree that contains an infinite number of copies of itself

the opposite sense to create ever smaller copies. This property, called *self-similarity*, is vividly illustrated by the definition of τ as a continued fraction:

$$x = 1 + \cfrac{1}{1 + \cfrac{1}{1+\cfrac{1}{1+\cfrac{1}{1+\dots}}}} = 1 + \frac{1}{x}; \quad x = 1 + \tau$$

Truncation of the algorithm at finite steps generates a series of rational fractions that converges to $1 + \tau = \Phi$. The convergents are the numbers

$$1,\ 1+1=2,\ 1+\frac{1}{1+1}=\frac{3}{2},\ 1+\frac{1}{1+\frac{1}{1+1}}=\frac{5}{3},\ \text{etc.}$$

This sequence of fractions are given by

$$\left(\frac{m}{n}\right)_i = \frac{1}{1}, \frac{2}{1}, \frac{3}{2}, \frac{5}{3}, \frac{8}{5}, \frac{13}{8}, \dots,$$

a series of Fibonacci fractions, as before.

Self-similarity is illustrated particularly well as the property of a Fibonacci *tree*, which contains infinituple copies of itself as shown in Fig. 2. Each black dot represents the start of a fresh tree.

The self-similarity associated with the golden section is embodied geometrically in the spiral inscribed within a rectangle with sides in golden ratio. Removal of a square (gnomon) from such a golden rectangle leaves a smaller golden rectangle as residue. On continuing the process indefinitely, rectangles of diminishing size are created, as shown in Fig. 3.

To good approximation, the side lengths of successive gnomons decrease in line with the Fibonacci numbers, such that Fibonacci squares cover the composite golden rectangle, e.g.

$$1^2 + 1^2 + 2^2 + 3^2 + 5^2 + 8^2 + 13^2 = 13 \times (13 + 8) = 13 \times 21.$$

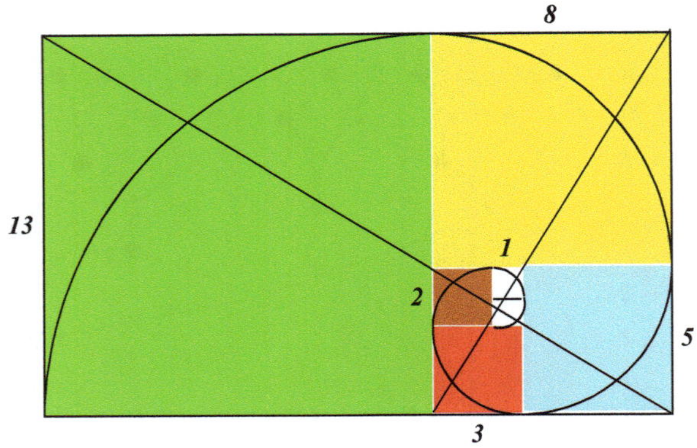

Fig. 3 A golden spiral inscribed in a golden rectangle defined by a Fibonacci sequence. The increasing size of successive squares is defined by the Fibonacci labels (n), and the ratio of their side lengths approaches τ as $n \rightarrow \infty$. The inscribed spiral approximates an equiangular logarithmic spiral [4]

Likewise, the first n rectangles in the Fibonacci sequence cover the largest gnomon for n odd, e.g.

$$(1\times 1)+(1\times 1)+(1\times 2)+(2\times 3)+(3\times 5)+(5\times 8)+(8\times 13)+(13\times 21) = 21^2$$

Circular segments inscribed between opposite vertices of the growing gnomons define a spiral which is approximated by the logarithmic spiral

$$r = a\,\mathrm{e}^{\theta \cot\varphi}, \qquad \varphi = 72.9^\circ, \qquad \cot\varphi = \tau/2\,.$$

3 The Periodic Function

Èmile de Chancourtois, co-discoverer of elemental periodicity, claimed [5] that

> ... the properties of the elements are the properties of numbers.

His claim was vindicated with the discovery of atomic number, but the theme remained undeveloped until it was conjectured by Plichta [6] that the electron configuration of atoms is mapped by the distribution of prime numbers. Based on the observation that all prime numbers >3 are of the type $6n \pm 1$, he defined a prime-number cross that intersects a display of natural numbers on a set of concentric circles with a period of 24. In Fig. 4, the construct is shown, rearranged as a number spiral. Noting that the numbers on each cycle add up to

$$\sigma(j+1) = \sum_{n=24j}^{24(j+1)} n = (2j+1)300\,,\ j = 0,1,2,\ldots$$

such that $\sigma_i = a, 3a, 5a, 7a\ldots\ \ (a = 300)$, these odd-number coefficients were likened to the degeneracy of spherical electronic shells with spectroscopic notation s, p, d, f.

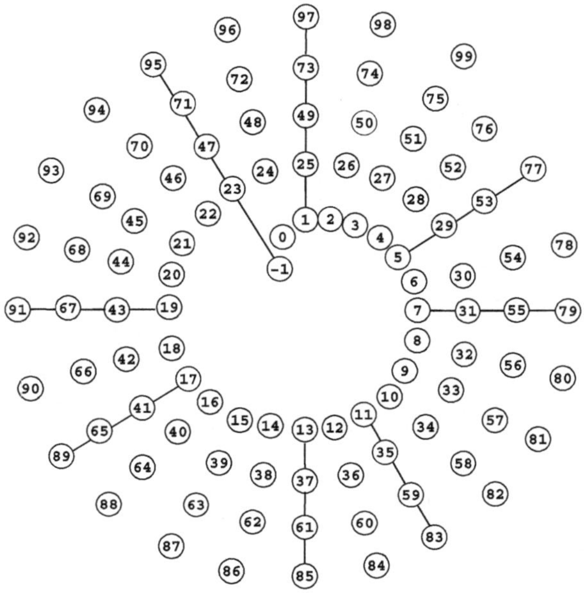

Fig. 4 The natural numbers arranged on a spiral with a period of 24. All prime numbers >3 and of the form $6n \pm 1$ occur on eight straight lines of the cross, which has been interpreted [6] to simulate the electronic structure of atoms

Reinterpretation of the sums as electron pairs over all stable nuclides suggests:

(a) A total[1] of 300 different nuclides
(b) 100 different elements
(c) Nuclide periodicity of 24
(d) Elemental periodicity of 8

This interpretation is supported [7] by analysis of the neutron imbalance of stable atomic species as a function of mass number, shown in Fig. 5. The region of nuclide stability is demarcated here by two zigzag lines with deflection points at common values of mass number A. Vertical hemlines through the deflection points divide the field into 11 segments of 24 nuclides each, in line with condition (c). This theme is developed in more detail in the paper on Atomic Structure in this volume. Defining neutron imbalance as either Z/N or $(N-Z)/Z$, the isotopes of each element, as shown in Fig. 6, map to either circular segments or straight lines that intersect where

$$\frac{Z}{N} = \frac{N-Z}{Z}, \text{ i.e. } \quad Z^2 + NZ - N^2 = 0,$$

with solutions of

$$Z = \frac{1}{2}N(1 \pm \sqrt{5}) \equiv \tau N.$$

[1] Three hundred isotopes of 100 elements with $Z/N = 1$ are synthesized by α-particle addition in massive stellar objects. In interstellar space radioactive decay terminates at 264 stable isotopes of 81 elements.

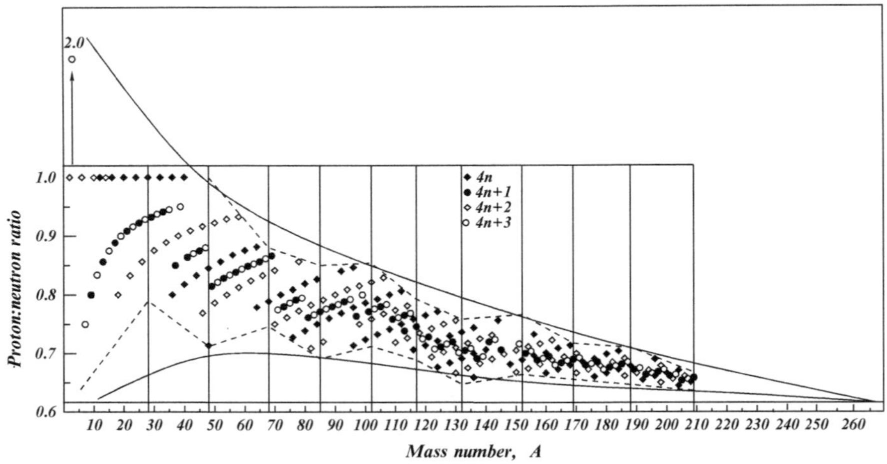

Fig. 5 Using half-life as criterion, the naturally occurring stable nuclides can be divided into four series of mass number $A(\text{mod}4) \equiv 0, 1, 2, 3$, i.e. two even series of 81 and two odd series of 51, considered as the product of α-particle addition [7]. A plot of the ratio Z/N vs A defines a converging field of stability, in line with the presumed periodicity of 24 (condition (c), see text)

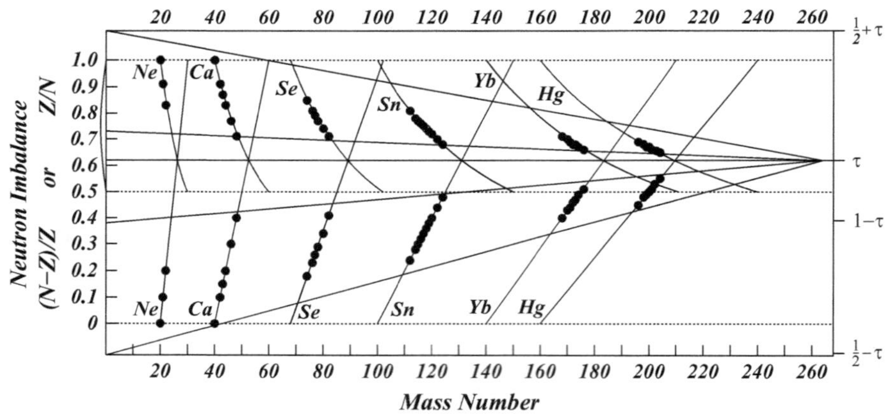

Fig. 6 Neutron imbalance of atomic nuclei is defined either by the ratio Z/N or the relative excess $(N - Z)/Z$. As functions of mass number, these quantities map the isotopes of a given element to respective circular segments and straight lines, the intersection of which defines the golden ratio

This result provides the exact value of the convergence limit of Z/N first identified by Harkins [8] as 0.62 according to the curves in Fig. 5.

To show that the periodic table of the elements is a subset of the more general nuclide periodicity, the data of Fig. 5 are replotted on axes of Z/N vs Z in Fig. 7.

In Fig. 7, the hemlines are no longer vertical, but still divide the field into 11 groups of 24. A remarkable feature of the diagram is the way in which the hemlines

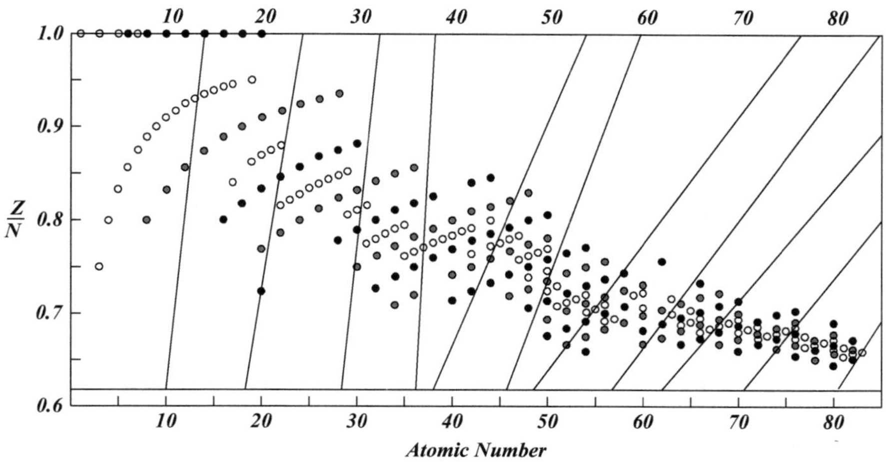

Fig. 7 The periodic distribution of stable nuclides as a function of atomic number. *Open circles* represent odd mass numbers and *filled circles* the two even mass-number series. The hemlines that define the nuclide periodicity of 24 are no longer parallel to the Z/N axis, and their points of intersection with the lines at $Z/N = \tau$ and 1 are of special importance in the definition of elemental periodicity as a subset of the nuclide periodic function

intersect the horizontal line $Z/N = \tau$ at points which, rounded off to the nearest integer, correspond with familiar values of atomic number that represent the closure of periodic subgroups:

$$10(2p), 18(3p), 28(3d), 36(4p), 38(5s), 46(4d), 48(5s), 56(6s), 62(4f : 6/8), 70(4f), 80(5d)$$

Extrapolated to $Z/N = 1$, the points of intersection represent an inverted periodic table:

$$14(4f), 24(3d), 32(1s), 38(3p), 54(5f), 60(4p), 76(5p), 84(6p), 94(5d), 100(7p).$$

The only known mechanism that could cause such an inversion is a state of extremely high pressure [9]. We conjecture that the unit ratio of Z/N coupled with high pressure describes ideal conditions for the build-up of 300 nuclides[2] of 100 elements by α-particle fusion in massive stellar objects, explaining conditions (a) and (b).

The cardinal points of intersection of the hemlines of Fig. 7 at $Z/N = 1.04$ are arranged symmetrically about $Z = 51$. On identification of the points $Z = 0$ and 102, a closed function is generated. Operating with the same element of mirror symmetry on the hemlines, two sets, characteristic of both nuclides and anti-nuclides, are generated as shown in Fig. 8.

[2]On release into interstellar space, radioactive decay results in the survival of only 264 stable nuclides as two sets of 81 with $A = 2n$ and two sets of 51 with odd A.

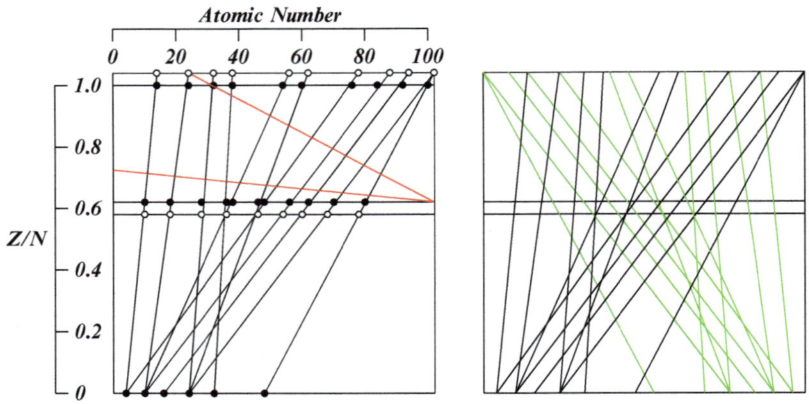

Fig. 8 Variability of the periodic table of the elements depends on space–time curvature as shown in the frame on the *left*. The triangular segment defines the field of stability. The symmetrical version on the *right* is conveniently mapped to the surface of a Möbius band as in Fig. 9, but resolution is only possible in 4D projective space

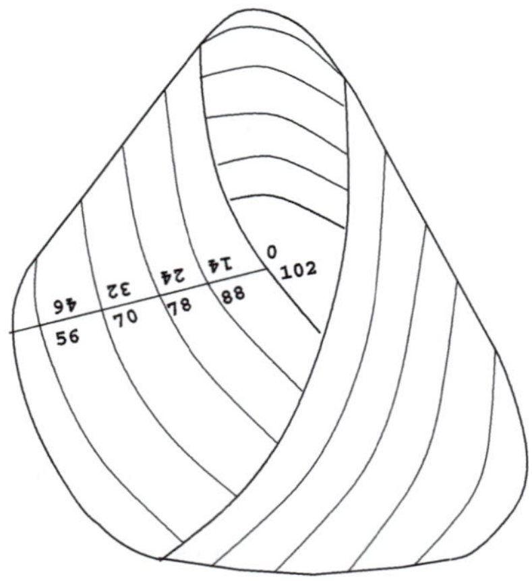

Fig. 9 Schematic diagram to illustrate the involuted nature of the periodic relationship between matter and antimatter

By following the hemlines from $Z/N = 0$ to 1.04 and back to zero, through the involution, a completely closed set is traced out, as shown in Fig. 9. Closed in four dimensions, the resulting topology defines real projective space. The implications of this construction on molecular shape is discussed in the final paper of this volume.

Like objects in the solar system (next section), the periodic table of the elements can also be rationalized by elementary number theory. As the ratio Z/N always represents a rational fraction, the pattern of Fig. 7 corresponds to some special ordering of rational fractions. The best-known order for the enumeration of rational fractions is known as a Farey sequence [10], which is generated by continued separate addition of numerators and denominators of adjacent fractions in the interval [0,1]:

$$\begin{array}{lllllllll}
\frac{0}{1} & & & & & & \frac{1}{1} & \mathcal{F}_1 \\[1ex]
\frac{0}{1} & & & \frac{1}{2} & & & \frac{1}{1} & \mathcal{F}_2 \\[1ex]
\frac{0}{1} & & \frac{1}{3} & \frac{1}{2} & \frac{2}{3} & & \frac{1}{1} & \mathcal{F}_3 \\[1ex]
\frac{0}{1} & \frac{1}{4} & \frac{1}{3} & \frac{1}{2} & \frac{2}{3} & \frac{3}{4} & \frac{1}{1} & \mathcal{F}_4 \\
. & . & . & . & . & . & . & \mathcal{F}_n
\end{array}$$

The variation of Z/N with Z, shown in Fig. 7, mirrors the variation of the infinite k-modular sets of Farey sequences, defined by

$$S_k = \frac{n}{n+k}, \quad (n,k) = 0,1,2,\ldots, \quad k = -\begin{vmatrix} h_i & h_{i+1} \\ k_i & k_{i+1} \end{vmatrix},$$

as a function of n. For simple Farey sequences, $k = 1$. For example,

$$\begin{aligned}
\{S_0\} &= \frac{1}{1} \quad \frac{1}{1} \quad \frac{1}{1} \quad \frac{1}{1} \cdots \quad \text{modulus } 0 \\
\{S_1\} &= \frac{0}{1} \quad \frac{1}{2} \quad \frac{2}{3} \quad \frac{3}{4} \quad \frac{4}{5} \cdots \quad \text{modulus } 1 \\
\{S_5\} &= \frac{0}{5} \quad \frac{1}{6} \quad \frac{2}{7} \quad \frac{3}{8} \quad \frac{4}{9} \quad \frac{1}{2} \cdots \quad \text{modulus } 5
\end{aligned}$$

Part of these sets is plotted in Fig. 10 with the converging stability limits shown.

Each point within the triangle of stability represents a potential nuclide. However, naturally occurring nuclides are limited to four series of mass numbers $A(\text{mod}4) \equiv 0 \rightarrow 3$, interpreted to correspond to a process of nucleogenesis based on the fusion of α-particles to yield two even series of 81 members each and two odd series with 51 members [7]. Significantly, by this procedure all stable nuclides are identified correctly, except for a few $\alpha-$unstable ones. Isotopes of elements 43 and 61 are excluded naturally. For $n = Z$, isotopes of the same element are mapped to the same vertical line.

More simply, a plot of the unimodular Farey sequence

$$\mathcal{F}_4 = \left\{ \frac{0}{1} \, \frac{1}{4} \, \frac{1}{3} \, \frac{1}{2} \, \frac{2}{3} \, \frac{3}{4} \, \frac{1}{1} \right\}$$

as Ford circles [11], directly represents the periodic table as in Fig. 11.

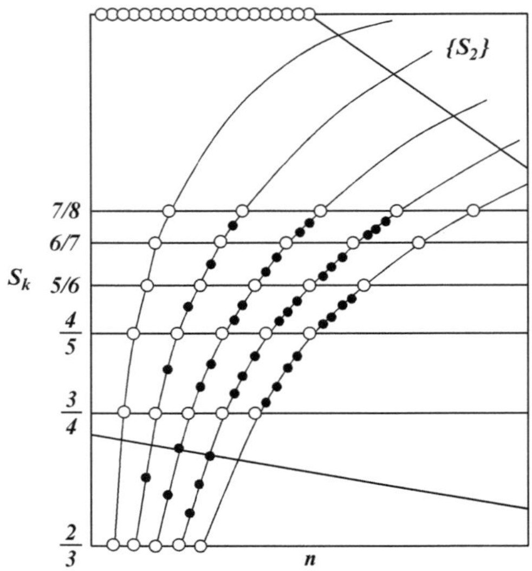

Fig. 10 A plot of k-modular Farey sequences as a function of the natural numbers defines a set of infinite festoons that resembles the arrangement of nuclides in Figs. 5 and 7. The segment, obtained as a subset defined by limiting Fibonacci fractions that converge from 1 to τ and subject to the condition $A(\text{mod}4) = 0 \rightarrow 3$, corresponds to the observed field of nuclide stability

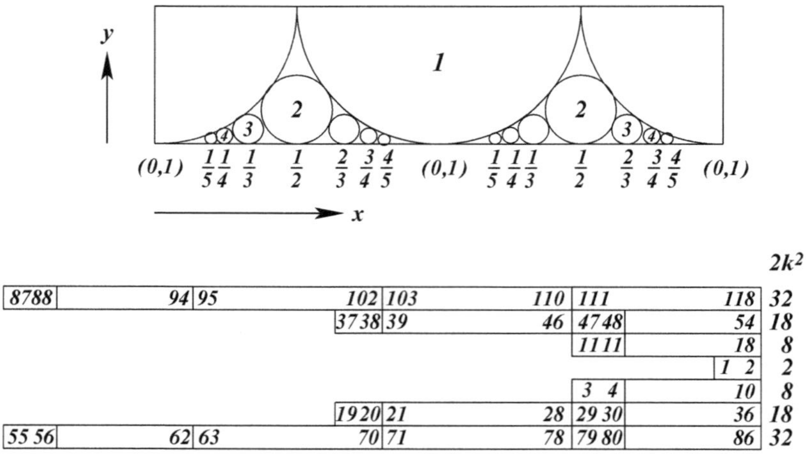

Fig. 11 Mapping of the periodic table of the elements as the reciprocal radii of the $\mathscr{F}_4$ unimodular Ford circles

As discussed in the paper on Atomic Structure, touching Ford circles have radii and y-coordinates of $1/2k_i^2$ and x-coordinates of h_i/k_i. The resulting map of $\mathscr{F}_4$ converts into the periodic table through the reciprocal radii of the numbered circles. Condition (d) is clearly implied.

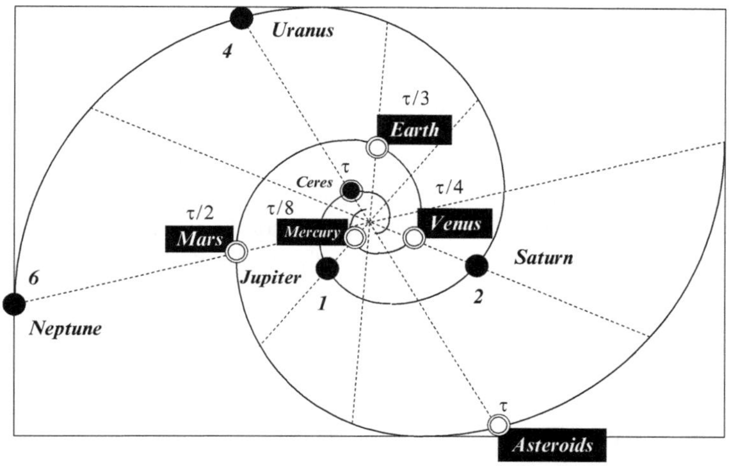

Fig. 12 Simulation of planetary orbits by golden-spiral optimization. With the mean orbital radius of Jupiter as unit, the outer planets are on orbits defined by integral multiples thereof. On the same scale, the asteroid belt is at a distance τ from the sun and the inner planets have orbital radii of τ/n. For clarity, the inner planets are shown on a larger self-similar scale

It all hangs together. To account for such consilience, Plichta [6] conjectured that numbers have real existence in the same sense as space and time. A more conservative interpretation would link numbers, through the golden ratio, to the curvature of space–time. A common inference is that the appearance of numbers as a manifestation of the periodicity of atomic matter is due to a spherical wave structure of the atom. A decisive argument is that the full symmetry, implied by the golden ratio, incorporates both matter and antimatter as a closed periodic function with involution, as in Fig. 9, in line with projective space–time structure.

4 Commensurability in the Solar System

Like the golden ratio, golden spirals give an exact description of a diversity of natural phenomena such as the shape of nautilus shells, tropical hurricanes and spiral galaxies [1]. Golden spirals have the property of self-similarity, or invariance with respect to scale transformation, in which small parts of a structure have geometrical properties that resemble the whole structure or large parts thereof. Numerically, both golden ratio and golden spiral are described by the convergence of Fibonacci series. The convergence properties have been put to scientific use in the theoretical reconstruction of atomic periodicity [7] and the solution of optimization problems in engineering [12]. As a significant demonstration, we show that the distribution of all matter in the solar system is correctly predicted by golden-spiral optimization. The assumption is that in a spiralling dust cloud, matter accumulates at specific points along the spiral, specified by a *convergence angle*. The result for a convergence angle of $(180/5)^{\circ}$ is shown in Fig. 12.

The mean orbits of all planets, including Ceres, the largest asteroid, are correctly predicted [13] by the relative distances from the spiral center. With the orbital radii expressed as rational fractions, a quantized distribution of major planets, as numbered, is revealed. On this scale the orbit of Ceres measures τ and those of the inner planets are rational fractions of the golden ratio. The same pattern was shown to repeat itself for the orbital motion of planetary moons and rings.

From this observation we infer self-similarity on a cosmic scale, from atoms to galaxies, which implies the same numerical basis for the atomic models of Nagaoka and Bohr, assumed to be self-similar with the rings of Saturn and the planets, respectively. In principle, the periodic accumulation of extranuclear electron density on an atom could also be optimized by specifying an appropriate convergence angle.

5 Atomic Structure

The Ford circles that represent the Farey sequence of order 4 represent the periodic table of the elements in complete detail. In particular, they predict the appearance of electron shells with $n = 1, 6$, consisting of 2, 8, 8, 18, 18 and 32 electrons, in this order.

This arrangement is ideally suited to optimization by a golden spiral. A chemically meaningful solution results by stipulating a divergence angle of $4\pi/(2n-1)$, as explained in the paper on Atomic Structure in this volume (see p. 71). The resulting distribution has extrema at integral distances of n^2 from the nucleus. If we assume that an electron cannot approach the nucleus more closely than the Bohr radius of a_0, the distribution can be considered as a spherical standing wave with zero surface at a radius a_0 and nodal surfaces at all n^2a_0. With the electron count known from the periodic function, the mean electron density within each shell follows immediately. The predicted distribution scales directly to the Thomas–Fermi statistical model of the atom, and the periodicity for individual atoms agrees with the Hartree–Fock results, see Fig. 13.

6 Electronegativity

The ground-state electron configuration of an atom is of limited value as a predictor of chemical affinity since most chemical reactions result from interaction between atoms in an activated valence state. The valence state is commonly considered to be reached by thermal activation, which gives rise to high-energy atomic collisions. Activation by uniform compression of an atom, which is not a feasible laboratory procedure, is more convenient for mathematical simulation [9, 14]. Computerized compression is done by Hartree–Fock simulation with modified boundary conditions that restrict wave functions to a finite sphere [14]. All electronic energies move to higher levels as the limiting radius is reduced, until a single electron reaches the

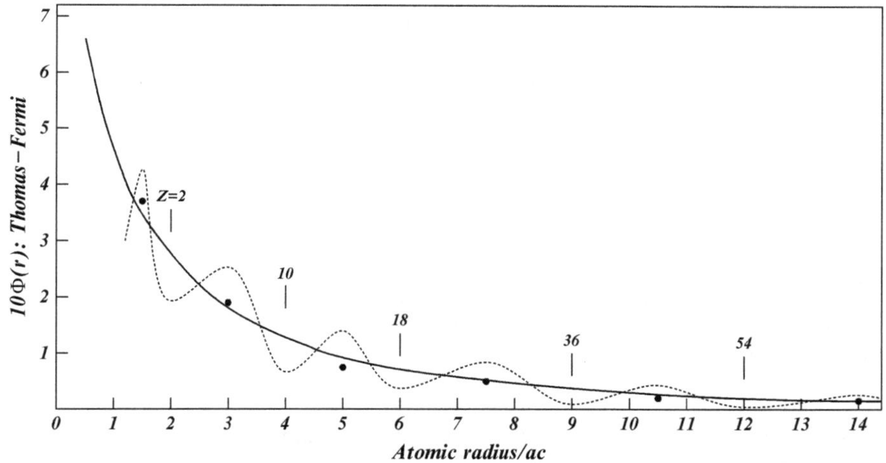

Fig. 13 Simulation of Thomas–Fermi and Hartree–Fock electron densities for unit atoms. The calculated points are those predicted by golden-spiral optimization and scaled to match the Thomas–Fermi curve, shown as a *solid line*. The *stippled curve* simulates the HF result

ionization limit. At this point, the valence electron, decoupled from interaction with the nucleus, is confined to a sphere of characteristic radius r_0, interpreted as the *ionization radius*. This is a standard problem in free-electron quantum theory which calculates the energy of an electron, which is uniformly spread across the confining sphere, as [15]

$$E_g = \frac{h^2}{8mr_0^2}.$$

When the valence electron reaches the ionization limit, its potential energy with respect to the nucleus goes to zero, and in uniform distribution, it has no kinetic energy. The calculated confinement energy (E_g) can therefore only represent quantum potential energy, defined as [16]

$$V_q = -\frac{\hbar\nabla^2 R}{2mR}.$$

This equation is solved by E_g, on the basis of which the valence electron is now interpreted to have reached an activated state at a chemical potential of E_g. This state defines *electronegativity*. To be in line with traditional practice, it is formally defined as $\chi = \sqrt{E_g}$, with r_0 in Å units and E_g in eV [17].

Starting from the numerically optimized valence density in a ground-state valence shell, the radius of an equivalent sphere, which accommodates this total density at a uniform level, is readily calculated by simple geometry. The correspondence with Hartree–Fock values of ionization radius is almost exact, with the added advantage of higher accuracy for the chemically important second period elements where the HF results are notoriously unreliable [14, 18].

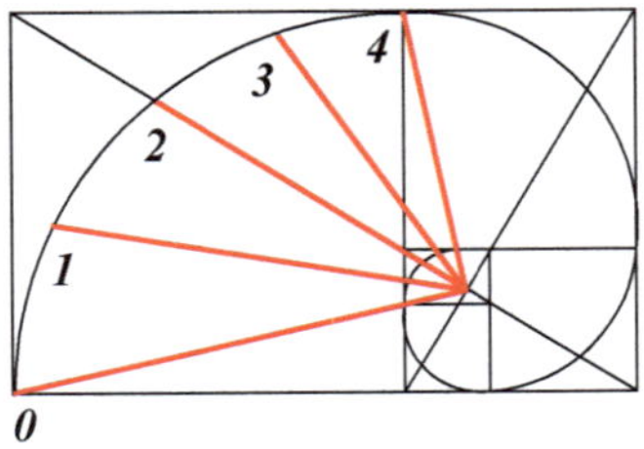

Fig. 14 Simulation of integer bond orders on a golden spiral. The dimensionless distances $d' = 1$ and τ, which represent zero and fourth order, respectively, must, by definition, be separated by a convergence angle of 90° on a golden spiral. Convergence angles for intermediate integer and half-integer orders follow directly

By the use of elementary number theory, to simulate uniform distribution of a valence electron over the ionization sphere, a complete set of ionization radii and electronegativities is now available for the simulation of a whole range of chemical properties as described in the papers to follow.

7 Bond Order

On looking for a relationship between ionization radius and the chemistry of homonuclear covalent interaction, the classification into single and multiple bonds[3] is followed as a first approximation. An immediate observation, valid for most single bonds, is a constant value of the dimensionless distance

$$d' = d/r_0 = 0.869,$$

where d is the experimentally known covalent interatomic distance. Notable exceptions occur for F–F, O–O and I–I with $d' = 0.935$. A similar trend is observed for most double and triple bonds. For the smaller number of quadruple dimetal bonds $d' = \tau$, to good approximation. Assuming zero-order interaction to occur at $d = r_0$, it follows that d' converges from unity to τ with increasing bond order, corresponding to a divergence angle of $\pi/2$ on a golden spiral.

Divergence angles of $\pi/8$ and $\pi/16$ for integer and half-integer bond orders are implied and shown in Fig. 14. This solution corresponds with the empirical values derived before, i.e.,

b	0	$\frac{1}{2}$	1	$1\frac{1}{2}$	2	$2\frac{1}{2}$	3	$3\frac{1}{2}$	4
d'	1.0	0.935	0.869	0.804	0.764	0.724	0.683	0.658	0.618

[3]Bond order, assumed to stipulate the number of electron pairs in covalent interaction, although a poor measure of bond strength [19–21], is a convenient general working model.

The corresponding numerical solution defines bond order, b, by the equation

$$d' = j_b \tau^n$$

where both n and the covariant integer coefficient j_b depend in a simple way on Fibonacci numbers. This procedure is described more fully in the paper on Bond Order in this volume.

The quantized variation of bond order may be rationalized by viewing overlapping charge spheres as spherical standing waves. These waves interfere constructively at specific interatomic distances that depend on wavelength. Destructive interference that occurs at intermediate distances tends to destabilize the interaction and to prevent continuous variation of bond order. Distortion of the interference pattern requires work, as measured by bond-stretching force constants.

8 Covalent Interaction

With the relationship between ionization radius and bond order in hand, the calculation of covalent interaction parameters becomes an almost trivial exercise. The common volume, ε, between overlapping spheres of radius r_0 at characteristic separations d' for given bond order, and considered proportional to dissociation energy, varies in a quantized fashion similar to d'. This allows definition of a dimensionless dissociation energy $D' = Dr_0/K$, as explained in the paper on Covalent Interaction (see p. 93), K is a dimensional constant. Noting the connection of ε with spherical volume, one looks for a dependence of the type

$$\frac{D_x r_0}{K} = D' \propto r_0^3.$$

First-order homonuclear interactions are seen to obey the rule, $D_x = K r_0^2 \tau^n$, where n correlates positively with bond order.

By taking electronegativity differences into account, dissociation energies for heteronuclear interactions are calculated as

$$D_c = K r_0^3(1)/r_0(2); \quad r_0(1) > r_0(2).$$

Exhaustive testing has shown the formula to hold for all heteronuclear interactions of any order.

Resistance against change of bond order is measured as a harmonic stretching force constant. It depends on the relative energies of adjacent bond orders and on the slope of the linear curve that describes continuous change of bond order. In general,

$$\frac{1}{2}k_r = \frac{\Delta D'}{(\Delta d')^2}.$$

Estimation of $\Delta D'$ is facilitated by the special property of the golden ratio, $\tau^n - \tau^{n+1} = \tau^{n+2} = \tau^+$, in shorthand notation. In the common units of N cm^{-1}, the general expression becomes

$$k_r = \frac{4.615\tau^+ s}{(\Delta d')^2 r_0(1) \cdot r_0(2)}.$$

The formula can be demonstrated convincingly, applied to well-studied series of diatomic molecules. Individual bonds in larger molecules are less well described, but MM force constants derived by this method are in good agreement with other general parameters [22].

Exploiting the possibility of representing polarization effects in diatomic molecules, many dipole moments, especially for diatomics such as the alkali halides, can be simulated with a high degree of confidence. A general lack of experimental values to serve as empirical guidance has so far prevented the development of the method to its full capacity.

9 The Physical Meaning

Based on elementary number theory, we have arrived at a computational scheme that works unexpectedly well for the simulation of chemical phenomena. The problem is to find a plausible interpretation to connect the abstract model with the concrete. The statistical scheme of probability densities and point particles adopted in physics does not provide a satisfactory answer in this case. We are forced to enquire more deeply into the fundamental nature of matter.

The only fundamental theory that considers the genesis of matter is the theory of general relativity. It is formulated in 4D space–time as a set of field equations,

$$G_{\mu,\nu} = kT_{\mu,\nu}, \quad \mu, \nu = 0, 3$$

that balances space–time curvature against the matter-energy content of the system. The implication of this relationship is that the appearance of matter is unequivocally linked to the curvature of space–time. In particular, matter cannot exist in Euclidean space [23].

For convenience, humans consider their living space to be Euclidean, and even the most advanced cosmologies are still formulated in such terms. In order to operate in Euclidean space, it is necessary to separate the mathematically equivalent variables of Einstein's equation into a universal time variable and the three familiar variables of coordinate space. This operation destroys the 4D field equations and no longer provides any insight into the nature of matter. For this reason, the existence of matter is added to physical theory as an ad hoc postulate—the prescription of classical Newtonian mechanics.

Quantum mechanics follows the same prescription, by separating the 4D potential function, formulated as

$$\Box^2 \Phi = 0 \quad \text{into the wave equation} \quad \left(\nabla^2 - \frac{1}{c^2}\frac{\partial^2}{\partial t^2}\right)\Phi = 0.$$

The suspicion that this separation of variables, although mathematically sound, leads to a less than perfect description of quantum systems is confirmed [24] by statistical testing of the seminal equation that relates the frequency of the energy radiated or absorbed by a H atom to the integers in the Rydberg formula

$$\bar{\nu} = R\left(\frac{1}{n^2} - \frac{1}{k^2}\right).$$

This is the equation of a straight line, which, for the Lyman series with $n = 1$, reduces to

$$\bar{\nu} = -R/k^2 + R$$
$$(y = mx + c)$$

A plot of $\bar{\nu}$ vs $-1/k^2$ should be linear, with equal slope and intercept. The most reliable and accurate data available fail this test statistically [24]. The discrepancy is not large but significant and reminiscent of the spectroscopic red shift measured in galactic light. The common origin of these discrepancies cannot be a Doppler effect and most likely is due to space–time curvature.

In the 4D equation, space and time coordinates are inextricably entangled. Its mathematical solutions are hypercomplex functions, or quaternions, without a commutative algebra. Quaternions are used to describe what is known as spherical rotation, also called the spin function, and the complex rotation known as the Lorentz transformation of special relativity.

The separated 3D wave equation can no longer describe any of these fundamental rotations. For this reason, the discovery of electron spin necessitated its introduction into quantum theory as another ad hoc postulate. The anomalous consequence is the unphysical situation of a point particle with spin. Only part of the spin function survives solution of the 3D wave equation, in the form of a complex variable interpreted as orbital angular momentum. In the so-called p_z-state, it has the peculiar property of non-zero orbital angular momentum with zero component in the direction of an applied magnetic field. This is the price to pay for elimination of spherical rotation.

Despite this anomaly, the complex angular-momentum function can be used to rationalize many features of stereochemistry. It represents the only vector quantity in wave mechanics which could interact with an applied magnetic field. It accounts for the so-called *Faraday effect*, which is a property of achiral molecules in an applied magnetic field to rotate the plane of polarised light. It is easy to demonstrate that atomic orbital angular-momentum vectors always line up antiparallel in diatomic interaction and continue to do so in more complicated symmetrical atomic assemblies on molecule formation. It is only in chiral molecules, without reflection or inversion symmetry, that total quenching of orbital angular momentum is no longer possible. The residual orbital angular momentum with its associated magnetic moment is responsible for optical activity in such cases.

This tendency to quench orbital angular momentum may be used to good effect to predict the mutual orientation of sub-molecular fragments (radicals) on forming a molecule. The symmetry of substituted methanes and other small molecules has been successfully analyzed by this method [15, 25]. Where quenching requires a specific mutual orientation in 3D, the interaction exhibits steric rigidity. By systematic use of these results, it is possible to predict the 3D structure of complex molecules, without allowing for the effects of non-bonded and torsional interactions. The resulting structure is suitable as trial input for optimization by MM [22].

These positive results are not without exception. Immediately obvious are the anomalous optical effects observed in a homochiral series such as the biologically active amino acids. Despite their uniform chirality,[4] only about half of them rotate an optical plane of polarization in the same sense. The orbital angular momentum of 3D wave mechanics is clearly inadequate as an explanation of optical activity in such detail. For a deeper understanding of the phenomenon, it is necessary to reconsider the effect of 4D molecular symmetry.

The discovery of quantum mechanics was seen as a dramatic departure from classical theory because of the unforeseen appearance of complex functions and dynamic variables that do not commute. These effects gave rise to the lore of quantum theory as an outlandish mystery that defies comprehension. In our view, this is a valid assessment only in so far as human beings have become evolutionary conditioned to interpret the world as strictly three dimensional. The discovery of a 4D world in special relativity has not been properly digested as yet, because all macroscopic structures are three dimensional. Or, more likely, minor discrepancies between 4D reality and its 3D projection are simply ignored. In the atomic and molecular domains, where events depend more directly on 4D potential balance, projection into 3D creates a misleading image of reality. We argue this point on the basis of different perceptions of chirality in 3D and 2D, respectively.

As shown in Fig. 15, inversion of a chiral tetrahedron, with the base *ABC* in a fixed 2D plane, changes the 3D chirality but appears as a simple rotation in 2D. We anticipate similar discrepancies between 4D symmetry and its projection into 3D as responsible for the irregular variation of optical activity as a function of 3D chirality.

The appearance of non-commuting quantum variables can now also be traced back to the non-commutative algebra of 4D hypercomplex functions. On projection into 3D by the separation of space and time variables, the quaternion variables are reduced to complex functions that characterize orbital angular momentum, but the commutation properties remain. Not appreciating the essence of complex wave

[4] For the sake of simplicity, we do not consider degrees of chirality as distinguished e.g. by chirality functions [26–28].

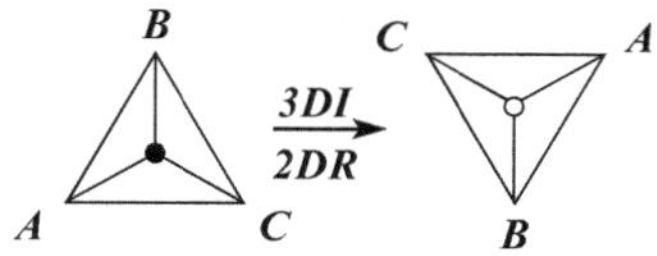

Fig. 15 Diagram to show that three-dimensional inversion (3DI) could be mistaken as rotation (2DR) in two dimensions

functions, an unfortunate tradition to reduce them to real functions has developed in quantum chemistry. These real orbitals and basis sets only have classical meaning. Relations in the complex plane, which describe orbital angular momentum, are reduced to harmonic oscillations on a real line.

Topologists describe the projection from four to three dimensions in terms of an underlying 4D curved space–time and a Euclidean space, tangent to the 4D manifold, with a universal time coordinate. The only 4D sphere with a continuous group structure is $\mathbb{S}^3$, and the space of antipodal points on $\mathbb{S}^3$ is known as *projective space* and denoted by $P_3(\mathbb{R})$ [29]. A section through projective space is a Möbius band—a 2D construct which cannot be embedded in 2D space. Correspondingly, a Möbius band, closed on all sides, cannot be embedded in 3D space. Projective space, although hard to visualize, is the physically most likely structure of the underlying 4D curved space–time.

The previous statement is based on the results of reformulating the theory of general relativity in projective space, as a model of unified electromagnetic and gravitational fields [30]. On transformation of the electromagnetic part into tangent space, the relationship between 3D and 4D potentials is of the form [31]

$$\left(\frac{e}{mc^2}\right) V_j = \frac{\sqrt{5}}{2}\varphi_j = \left(\tau + \frac{1}{2}\right)\varphi_j \, .$$

Here is the first inkling about the universal importance of the golden ratio. If it measures the relationship between underlying space and tangent space, it is not surprising for it to show up in the apparent structure of so many objects from atoms to galaxies. It could even be interpreted as a measure of space–time curvature.

The relationship between space–time curvature and the golden ratio amounts to little more than a suspicion, based on the factor that converts the electromagnetic field from projective space into an affine theory. To account for the ubiquitous appearance of golden spirals in self-similar chemical and cosmic structures, we may look for a related factor in the construction of a golden spiral from the series of gnomonic circular segments as in Fig. 3. The ratio between arc length, $\pi r/2$, and cord, $\sqrt{2}r$, i.e. between curve and tangent, $\pi/(2\sqrt{2}) = 1.111 \simeq \sqrt{5}/2$, confirms the suspicion. If the curvature of space–time depends on the golden ratio, it is no longer surprising to find that structures of all sizes are self-similar and conditioned by τ. To summarize, the golden ratio features in the packing of nucleons [7], the electronic structure of atoms, the details of chemical interaction, the periodicity of atomic matter [2], the structure of nanoparticles [3], botanical phyllotaxis, planetary and solar systems [13], spiral galaxies and the large-scale structure of the universe [23].

It is reasonable to expect that any growing structure should follow this curvature and for 3D observers to find the imprint of the golden ratio everywhere, also in the structure of molecules. It would not be surprising to find the golden ratio among the topological features of macromolecules or even reflected in the torsion angles within smaller molecules [32].

We are reaching the conclusion that the modeling of molecular properties by number theoretic golden parameters, rather than a meaningless coincidence, may well be the fundamentally most appropriate procedure. At this stage, the simulation of the important internal molecular parameters, with the exception of torsion angles, can be undertaken with confidence. The final objective would be structure optimization by MM, using a force field based entirely on number theory.

10 Molecular Mechanics

Molecular modeling by minimization of the steric energy starts from an assumed structure in which the rigid bonds are replaced by flexible springs of characteristic lengths and obey Hooke's law [33].

In the empirical approach, a characteristic bond length and a matching force constant are assigned to reproduce an observed interatomic distance on energy minimization. In the number theory approach [22], an ideal bond length, free of strain, is calculated directly as a function of ionization radii and bond order. For smaller, chemically important atoms, it is necessary to allow for the distortion of atomic sphericity by first-neighbor ligands. Calculation of the stretching force constant requires, in addition, an estimate of the diatomic dissociation energies at different bond orders, which are also obtained as functions of ionization radii and the golden ratio.

Second-neighbor interactions are optimized empirically from a characteristic valence angle and a matching angle-bending force constant. In the number theory approach, 1,3-interactions are calculated as a low-order interaction with an associated stretching force constant, readily converted into an angle-bending force constant as needed. The interaction between more distant neighbors are considered as a weak bond with a minimum at the characteristic van der Waals separation. In number theory, these are zero-order interactions. The parameters generated by number theory would therefore be ideally suited for use with a central force field [34] in which two-body forces between atoms are used exclusively, in place of bond angles and torsion angles.

Torsional interactions are empirically modeled with a periodic function based on local geometry around the central bond. This problem has not been solved in number theory. Neither have any of the other parameters such as cross-term interactions, electrostatic interactions, out-of-plane deformation and the like been considered for number theory simulation. Most of these are not needed when using a central force field. However, the purpose of the number theory analysis is not to produce another force field, but rather to demonstrate that the MM approach is fundamentally sound.

An important contribution that number theory can make to MM simulations is to also provide force-field parameters for interactions that involve metal ions, even if these are subject to further empirical refinement [22].

11 Chemical Matter

As in wave mechanics, the simulation of chemical phenomena by number theory is characterized by the appearance of integers, in this case associated with chemical structures and transformations. An obvious conclusion is that the elementary units of matter should be viewed as wave structures rather than point particles, which is consistent with the first appearance of matter in curved space–time. Even 3D wave packets behave in a manner convincingly like ponderable matter and rationalize the equivalence of mass and energy in a natural way. There is no compelling reason why this simple model should be concealed with the notion of wave/particle duality and more so on realizing that the wave-like space–time distortions are strictly 4D structures. In response to environmental pressure, an electronic wave packet can shrink to the effective size of an elementary particle or increase to enfold a proton as a spherical standing wave.

A wave structure of the electron, which is routinely verified by electron diffraction, facilitates the understanding of atomic structure and concepts like bond order. It provides a logical explanation of the photoelectric effect as an interaction between waves, doing away with photons and wavicles at the same time.

Interacting elementary wave packets are expected to coalesce into larger wave packets. All extranuclear electrons on an atom therefore together constitute a single spherical standing wave with internal structure, commensurate with a logarithmic optimization pattern. In the activated valence state, the central core of the wave packet is compressed into a miniscule sphere, compared to the valence shell which dominates the extranuclear space up to the ionization radius.

When atoms interact, these standing waves interfere to generate an interference pattern that determines the molecular charge distribution. Closer interaction happens stepwise in stages, described in chemical terminology as increasing bond order. Charge densities recorded in X-ray crystallographic studies clearly resemble such a distribution.

Acknowledgements Financial support by the Alexander von Humboldt Foundation is gratefully acknowledged. We acknowledge the input of Casper Schutte during many informative discussions.

References

1. Livio M (2002) The golden ratio. Review, London
2. Boeyens JCA (2003) J Radioanal Nucl Chem 259:33
3. Li C, Zhang X, Cao Z (2005) Science 309:909
4. Coxeter HSM (1989) Introduction to geometry, 2nd edn. Wiley, New York
5. de Chancourtois AEB (1889) Nature 41:186
6. Plichta P (1998) God's secret formula. Element Books, Boston
7. Boeyens JCA, Levendis DC (2008) Number theory and the periodicity of matter. Springer Dordrecht, Heidelberg, London, New York
8. Harkins WD (1931) Phys Rev 38:1270
9. Goldman S, Joslin C (1992) J Phys Chem 96:6021
10. Hardy GH, Wright EM (1979) An introduction to the theory of numbers, 5th edn. University Press, Oxford
11. Ford LR Sr (1938) Am Math Month 45:586
12. Boltyanskii VG (1971) Mathematical methods of optical control. Translated from the Russian by Trirogoff KN and Tarnove I. Holt, Reinhart and Winston, New York
13. Boeyens JCA (2009) Phys Essays 22:493
14. Boeyens JCA (1994) J Chem Soc Faraday Trans 90:3377
15. Boeyens JCA (2008) Chemistry from first principles. Springer.com
16. Bohm D (1952) Phys Rev 85:166
17. Boeyens JCA (2008) Z Naturforsch 63b:199
18. Herman F, Skillman S (1963) Atomic structure calculations. Prentice-Hall, Upper Saddle River
19. Boeyens JCA (1982) J Crystallogr Spectr Res 12:245
20. Kutzelnigg W (1973) Angew Chem 85:551
21. Pauling L (1960) The nature of the chemical bond. Cornell University Press, Ithaca
22. Comba P, Boeyens JCA (2013) Struct Bond 148:137
23. Boeyens JCA (2010) Chemical cosmology. Springer, Dordrecht, Heidelberg, London, New York
24. Schutte CJH (2013) Struct Bond 148:49
25. Boeyens JCA (2003) New theories for chemistry. Elsevier, Amsterdam
26. Derflinger G (1991) In: Janoschek R (ed) Chirality—from weak bosons to the α-helix. Springer, Dordrecht, Heidelberg, London, New York
27. Janoschek R (ed) (1991) Chirality—from weak bosons to the α-helix. Springer, Heidelberg
28. Ruch E (1972) Acc Chem Res 5:49
29. Stillwell J (1992) Geometry of surfaces. Springer, Dordrecht, Heidelberg, London, New York
30. Veblen O (1933) Projektive Relativitätstheorie. Springer, Berlin. English translation in Boeyens JCA (2010) Chemical cosmology. Springer, Dordrecht, Heidelberg, London, New York
31. Veblen O, Hoffman B (1930) Phys Rev 36:810
32. Boeyens JCA (2010) Int J Mol Sci 11:4267
33. Comba P, Hambley TW, Martin B (2009) Molecular modeling of inorganic compounds, 3rd completely revised and enlarged edition. Wiley-VCH, Weinheim
34. Saunders M, Jarret RM (1986) J Comp Chem 7:578

Struct Bond (2013) 148: 25–47
DOI: 10.1007/978-3-642-31977-8_2
© Springer-Verlag Berlin Heidelberg 2013

Chemistry in Four Dimensions

Jan C.A. Boeyens

Abstract Some chemical phenomena, awkward to rationalize, are argued to originate in the four-dimensional nature of matter in curved space–time. The problem is traced back to the separation of space and time variables in the analysis of four-dimensional events. Although mathematically sound, this operation is not physically valid. It destroys the essential non-classical entanglement of space and time, which is recognized in relativistic theory, but not in quantum mechanics. We show that without this approximation, the state functions of quantum theory have the same quaternion structure that describes Lorentz transformation and spin. Hypercomplex formulation of four-dimensional motion eliminates several bothersome concepts, such as wave–particle duality and probability density, by providing the logical basis for non-zero commutators in non-classical systems. It shows why chiral states are undefined in quantum theory and why many solid-state transitions appear to be sterically forbidden. A brief introduction to hypercomplex algebra is given as an Appendix A.

Keywords d'Alembertian · Dirac's equation · Harmonic function · Quaternion · Spin function

Contents

J.C.A. Boeyens (✉)
Unit for Advanced Scholarship, University of Pretoria, Lynnwood Road,
0002 Pretoria, South Africa
e-mail: jan.boeyens@up.ac.za

1 Introduction

Several generations of chemists have been conditioned to accept the notorious discrepancy between the theory and practice of chemistry as the unquestionable norm. Sterically forbidden molecular rearrangements and phase transformations are routinely reported without comment, and the flow of electronic particles, postulated to rationalize the course of chemical reactions, is never subjected to critical scrutiny. In reality, practising chemists design their experiments in terms of the nineteenth-century notions of chemical affinity, never adequately explained by twentieth-century theories. The innocent belief that quantum physics explains "all of chemistry" is, like the rest of quantum theory, obediently respected as just another of its deep inscrutable mysteries.

The reluctance to abandon dogmatic theory often results in the introduction of secondary ad hoc explanations to cover up any cracks in the theory, as they occur. A prime example occurs in the quantum theory of elemental periodicity. Based on the wave-mechanical ordering of electronic energy levels in an isolated H atom, a logical Aufbau procedure for many-electron atoms suggests itself. Elemental ordering in shells, consisting of $2n^2$ successive atoms, for integral $n > 0$, is predicted. This pattern breaks down at $n = 3$. In order to rescue the theory, a poorly explained effect ascribed to inter-electron repulsion is conjectured, without modification of the basic theory. Even if the only effect of the additional assumption was the splitting of energy sub-levels whereby, for instance, the $4s$ sub-level occurs at a lower energy than $3d$, it still fails to account for the observed periodicity. Instead of the expected $4s^{1\rightarrow 2}3d^{1\rightarrow 10}$, the sequence $4s^{1\rightarrow 2}3d^{1\rightarrow 8}(3d^{10}4s^{1\rightarrow 2})$ is observed. Contrary to Aufbau philosophy, the interpolated transition series is almost uniformly characterized by a $4s^2$, rather than $3d$, valence shell. Arbitrary new concepts, such as the degree of orbital penetration towards the nucleus, without any wave-mechanical basis, are invoked next, without improving the theory in any way. A more appropriate response would have been to admit failure of the H model, applied to non-hydrogen atoms.

The unjustified confidence in wave mechanics to account for elemental periodicity encouraged the belief in quantum theory to explain all chemistry, albeit with the aid of supercomputers—another unfulfilled dream. The cruellest reality for chemists has perhaps been the consistent failure to identify the quantum-mechanical basis of the three-dimensional structure of molecules, substantiated by thousands of independent observations, and despite massive computational efforts. This negative result provides a significant clue to account for the failure of quantum theory in the description of chemical systems.

One of the more successful devices to reconcile chemical behavior with quantum theory was the proposed definition of atomic orbitals to regulate the distribution of electrons on both atoms and molecules. A minor irritant in this application is the complex nature of the relevant wave functions that underlie the definition of atomic orbitals. As these complex functions invariably occur in orthogonal pairs, real functions can be constructed by suitable linear combinations of these pairs.

The same strategy is used in all quantum-chemical computations in order to avoid complex variables. This strategy comes at an exorbitant price.

A wave-mechanical model of the H atom describes an electron in terms of three quantum numbers. However, in order to account for atomic spectra, it is necessary to assume that the extranuclear electrons are not all concentrated at the lowest energy level, but distributed over several levels as stipulated by a fourth quantum number, postulated to represent a two-level spin system that obeys an exclusion principle. The strict consequence of this observation is that the orbitals of a threefold degenerate level must have the third quantum number with values of $m_l = -1, 0, 1$, which eliminates the possibility of three real functions (all $m_l = 0$). The simple conclusion is that any computational scheme that operates exclusively on real variables cannot be considered to be quantum mechanical, but rather as strictly classical.

The pioneers of quantum theory were unprepared for the discovery of two related, but unforeseen, new effects—the appearance of complex variables as an essential feature of the theory and pairs of conjugate variables that do not commute—a stark departure from classical mechanics. The appearance of complex variables is now readily appreciated, even in two-dimensional harmonic functions, defined by the equation:

$$\frac{\partial^2 \Phi}{\partial x^2} + \frac{\partial^2 \Phi}{\partial y^2} = 0\,.$$

One method to solve this equation is by assuming the potential function to be the product of two variable functions of x and y, respectively, i.e.

$$\Phi = X \cdot Y$$

which on differentiation gives

$$\frac{1}{X}\frac{\partial^2 X}{\partial x^2} + \frac{1}{Y}\frac{\partial^2 Y}{\partial y^2} = 0. \tag{1}$$

The two terms are functions of independent variables and, in order to satisfy (1), must be independent of both variables and therefore equal to a constant, i.e.

$$\frac{1}{X}\frac{\partial^2 X}{\partial x^2} = k_x^2; \qquad \frac{1}{Y}\frac{\partial^2 Y}{\partial y^2} = k_y^2.$$

To satisfy the implied condition, $k_x^2 + k_y^2 = 0$, either $k_x = k_y = 0$ or $\mathrm{i}k_x = k_y$, such that

$$\Phi = c\mathrm{e}^{\pm k(x+\mathrm{i}y)},$$

which describes a rotation in complex notation, exactly as required by wave mechanics.

In three-dimensional formulation, the condition $k_x^2 + k_y^2 + k_z^2 = 0$ or $(A + \mathrm{i}B)^2$ is satisfied, without loss of generality on setting $k_z = 0$, $\mathrm{i}k_x = k_y$, to describe a threefold degenerate state by the magnetic quantum number $m_l = 0, \pm 1$. Equating all constants to zero, by the mathematical separation of the physically entangled x and y coordinates, not only avoids the use of complex functions, but also destroys the ability to describe the angular momentum of the system. The one-dimensional projection appears as harmonic oscillation, e.g.

$$\frac{\partial^2 X}{\partial x^2} = k^2 X, \quad X = a\mathrm{e}^{\pm kx},$$

which is an incomplete description of the complex rotation.

We have reached the uncomfortable conclusion that our trusted three-dimensional wave equation, interpreted as a projection from four-dimensional space–time, must have the same defects as a harmonic oscillator when being considered a model of complex rotation.

2 Four-Dimensional Motion

Motion in four dimensions proceeds against the gradient of a scalar function Φ:

$$\Box\Phi = \sum_{\mu=0}^{3} \frac{\partial \Phi}{\partial x_\mu} = \sum \partial \Phi_\mu \equiv \left(\frac{1}{\mathrm{i}c}\frac{\partial}{\partial t} + \frac{\partial}{\partial x} + \frac{\partial}{\partial y} + \frac{\partial}{\partial z}\right)\Phi$$

and the potential field is described by the d'Alembertian:

$$\Box^2\Phi = \left(\frac{\partial^2}{\partial x_0^2} + \frac{\partial^2}{\partial x_1^2} + \frac{\partial^2}{\partial x_2^2} + \frac{\partial^2}{\partial x_3^2}\right)\Phi = 0 \tag{2}$$

$$\equiv \left(\nabla^2 - \frac{1}{c^2}\frac{\partial^2}{\partial t^2}\right)\Phi = 0 \tag{2a}$$

Equation (2) is recognized as Laplace's equation in four dimensions and in alternative form is known as the wave equation in three-dimensional space.

The mathematical procedure of solving (2) is by treating Φ as the product function:

$$\Phi = X_0(x_0) \cdot X_1(x_1) \cdot X_2(x_2) \cdot X_3(x_3).$$

By substituting the four factors,

$$\frac{\partial^2 \Phi}{\partial x_0^2} = \frac{\partial^2 X_0}{\partial x_0^2}(X_1 X_2 X_3), \text{ etc.,}$$

into (2) and dividing by Φ,

$$\frac{1}{\Phi}\Box^2\Phi = \frac{1}{X_0}\frac{\partial^2\Phi}{\partial x_0^2} + \frac{1}{X_1}\frac{\partial^2\Phi}{\partial x_1^2} + \frac{1}{X_2}\frac{\partial^2\Phi}{\partial x_2^2} + \frac{1}{X_3}\frac{\partial^2\Phi}{\partial x_3^2} = 0\,. \tag{3}$$

Each term is a function of a single variable and (3) can only remain valid if each term is independent of all variables and equal to a constant. Each term can hence be written in the form of a one-dimensional Helmholtz equation:

$$\frac{\mathrm{d}^2X}{\mathrm{d}x^2} - k^2X = 0 \tag{4}$$

in which $X = a\mathrm{e}^{\pm kx}$. For $X = T(t)$, the equation

$$\frac{1}{c^2}\frac{\mathrm{d}^2T}{\mathrm{d}t^2} + k^2T = 0\,, \tag{4a}$$

writing $\omega = ck$, yields $T(t) = a\exp(\pm\mathrm{i}\omega t)$, in which ω is interpreted physically as an angular frequency.

Each term in (3) requires an arbitrary constant to generate the overall solution $\Phi = A\exp(k_0x_0 + k_1x_1 + k_2x_2 + k_3x_3)$. These four constants are not independent and must satisfy the general condition: $\sum k_\mu^2 = 0$, imposed by (3) and (4). The most general solution of $\Box^2\Phi = 0$ is therefore a function that depends equally on all four coordinates, rather than the product of four linear functions. The four-dimensional Laplacian $\Box^2\Phi = 0$ describes the state of potential balance along the curved Riemannian manifold. Minkowski space–time, tangent to the manifold, describes a local pseudo-Euclidean approximation. By separating space and time variables, the Laplacian reduces to a wave equation (2a) in Euclidean space. This equation is the basis of wave mechanics, which further separates space and time variables into space-like and time-like equations.

It is at this point that wave mechanics moves out of Minkowski space. Away from the assertion that space–time is characterized by the harmonics of a four-dimensional Laplacian, visualized as long-wavelength undulation, like a wave field in Minkowski space. In wave-mechanical approximation, the time and space variables (collectively represented as x) are separated by definition of the product function

$$u(x,t) = f(x)\cdot\mathrm{e}^{-\mathrm{i}\omega_0 t}$$

to yield

$$\nabla^2\psi + \left(\frac{\omega_0}{c}\right)^2\psi = 0 \tag{5}$$

and

$$\left(\nabla^2 + \frac{\mathrm{i}\omega_0}{c^2}\frac{\partial}{\partial t}\right)\Psi = 0\,. \tag{5a}$$

By de Broglie's postulate, the wave vector becomes

$$k = \frac{\omega_0}{c} = \frac{2\pi}{\lambda} = \frac{p}{\hbar},$$

whereby (5) and (5a) reduce to the familiar Schrödinger equations, which provide a good practical approximation within the locally perceived three-dimensional space, but the holistic entanglement of space–time coordinates is lost. Developed into a theory of atomic stability, the equilibrium condition, $\mathrm{d}E/\mathrm{d}r = 0$, defines the so-called stationary states of the hydrogen atom as a function of the total electronic energy:

$$E = \frac{p^2}{2m} - \frac{\mathrm{e}^2}{r} \quad \text{(in esu).}$$

Considered as a standing de Broglie wave of wavelength $n\lambda = 2\pi r_n$, the distance between proton and electron follows as

$$r_n = \frac{(n\hbar)^2}{m\mathrm{e}^2}, \quad \text{i.e.} \quad E_n = -\frac{\mathrm{e}^2}{2r} = -\frac{m\mathrm{e}^4}{2(nh)^2} = -\frac{Rhc}{n^2}.$$

This inverse square relationship, which assumes a spherically symmetrical standing electron wave, is the fundamental equation of atomic spectroscopy and non-relativistic wave mechanics. In four-dimensional space–time, especially in a non-zero gravitational field, the assumption is not strictly valid and the proportionality factor R_n may vary with n.

The mass variable is a strictly empirical assumption that only acquires meaning in non-Euclidean space–time on distortion of the Euclidean wave field defined by Eq. (2). The space-like Eq. (5), known as Schrödinger's time-independent equation, is not Lorentz invariant. It is satisfied by a non-local wave function which, in curved space, generates time-like matter-wave packets, characterized in terms of quantized energy and three-dimensional *orbital* angular momentum. The four-dimensional aspect of rotation, known as *spin*, is lost in the process and added on by assumption. For macroscopic systems, the wave-mechanical quantum condition $\hbar\omega = E - V$ is replaced by Newtonian particle mechanics, in which $E = \frac{1}{2}mv^2 + V$. This condition, in turn, breaks down as $v \to c$.

Wave mechanics and particle mechanics, formulated to describe motion in three-dimensional space, are both incomplete by their failure to account for spin and relativistic effects. The common defect in both formulations lies in the unphysical separation of space and time variables. The proper procedure requires hypercomplex solutions of (2), which describe motion in four-dimensional space–time.[1]

Although hypercomplex solution of (2) has never been achieved, a conditional covariant form of the wave equation was proposed by Dirac [1] on assuming spin

[1] For the benefit of those readers who are not familiar with hypercomplex numbers and quaternions, an elementary introduction is provided as an Appendix A to this chapter.

matrices as possible solutions. The procedure consisted of modifying Schrödinger's equation

$$\mathrm{i}\hbar\frac{\partial\Psi}{\partial t} = H\Psi\,,$$

which is linear in $\partial/\partial t$, by inserting the relativistic Hamiltonian

$$H = \left(p^2c^2 + m_0^2c^4\right)^{\frac{1}{2}}\,.$$

To ensure Lorentz invariance, the Hamiltonian should also be linear in space derivatives, $-\hbar\mathrm{i}p_k = \partial/\partial x_k$ $(k = 1, 2, 3)$, such that

$$H = \sum_{k=1}^{3}\alpha_k c p_k + \beta m_0 c^2, \quad \text{i.e.}$$

$$\mathrm{i}\hbar\left[\frac{\beta}{c}\frac{\partial}{\partial t} + \sum_{k=1}^{3}\beta\alpha_k\frac{\partial}{\partial x_k}\right]\Psi = m_0 c\Psi\,.$$

With $\gamma^0 = \beta$, $\gamma^k = \beta\alpha_k$, $x_0 = ct$, the equation

$$\mathrm{i}\hbar\left(\gamma^0\frac{\partial}{\partial x_0} + \gamma^1\frac{\partial}{\partial x_1} + \gamma^2\frac{\partial}{\partial x_2} + \gamma^3\frac{\partial}{\partial x_3}\right)\Psi = m_0 c\Psi\,,$$

is Lorentz invariant and commonly transformed, using units of $\hbar = c = 1$ and the Einstein summation convention, to read

$$\left(\mathrm{i}\gamma^\mu\partial_\mu - m\right) = 0.$$

These conditions are fulfilled by assuming coefficients of the form

$$\alpha_k = \begin{bmatrix} 0 & \sigma_k \\ \sigma_k & 0 \end{bmatrix}, \quad \beta = \begin{bmatrix} I & 0 \\ 0 & -I \end{bmatrix}$$

where σ_k are Pauli matrices and I is the 2×2 unit matrix. As before, the spin variables are therefore introduced empirically.

The most significant difference of Dirac's results from those of the non-relativistic Pauli equation is that the orbital angular momentum and spin of an electron in a central field are no longer separate constants of the motion. Only the components of $\mathbf{J} = \mathbf{L} + \mathbf{S}$ and $\mathbf{J}^2$, which commute with the Hamiltonian, emerge as conserved quantities [1]. Dirac's equation, extended to general relativity by the method of projective relativity [2], automatically ensures invariance with respect to gauge, coordinate and spinor transformations, but has never been solved in this form.

The surprising implication is that Dirac's equation does not allow of a self-consistent single-particle interpretation, although it has been used to calculate approximate relativistic corrections to the Schrödinger energy spectrum of hydrogen. The obvious reason is that a 4D point particle is without duration and hence undefined. An alternative description of elementary units of matter becomes unavoidable. Prompted by such observation, Dirac [3] re-examined the classical point model of the electron only to find that it has three-dimensional size, with an interior that allows superluminal signals. It all points at a wave structure with phase velocity $v_\phi > c$.

The equation $\square^2 \Phi = 0$ has unitary quaternion solutions of the form

$$\Phi = \mathrm{e}^{\theta(\mathrm{i}\alpha+\mathrm{j}\beta+\mathrm{k}\gamma)} = \cos\theta + \sin\theta(\mathrm{i}\alpha + \mathrm{j}\beta + \mathrm{k}\gamma)\,,$$

which represent the $SU(2)$ rotational Lie group. The four-dimensional eigenfunction is the proper spin function. With $\theta = \omega t$, it is the state function that describes total angular momentum $\mathbf{J} = \mathbf{L} + \mathbf{S}$. For $\mathrm{i}\alpha + \mathrm{j}\beta + \mathrm{k}\gamma = \mathrm{i}$, it describes spin only. Finally, with $\beta = \gamma = 0$, it defines spin in polar projection ($s = \pm\frac{1}{2}$), as shown in more detail below. Alternatively, Φ is the state function of an electromagnetic photon.

With $\cos\theta = 0$, Φ could represent the space-like state function of electric charge, known as the angular-momentum function, described by the quantum number l in Schrödinger notation, also known as the three-dimensional spherical harmonics.

The most general rotation of a four-vector represented by $q = w + \mathrm{i}x + \mathrm{j}y + \mathrm{k}z$ is described by the equation

$$q' = \mathrm{e}^{\theta(\mathrm{i}\alpha+\mathrm{j}\beta+\mathrm{k}\gamma)} q \mathrm{e}^{-\varphi(\mathrm{i}\lambda+\mathrm{j}\mu+\mathrm{k}\nu)}$$

where $\alpha^2 + \beta^2 + \gamma^2 = \lambda^2 + \mu^2 + \nu^2 = 1$. When both θ and φ are imaginary angles, the transformation corresponds to the Lorentz transformation [4, 5]. This is mathematical proof that quantum theory and special relativity originate from the same basis, as the most general solution of (2) must obviously be a quaternion function.

3 Spherical Rotation and Spin

The quaternion that describes rotation through θ about the x-axis follows as

$$\mathrm{e}^{\mathrm{i}(\theta/2)} = \cos(\theta/2) + \mathrm{i}\sin(\theta/2)\,.$$

This hypercomplex number is given in matrix form by

$$q = \begin{pmatrix} 0 & \mathrm{e}^{-\mathrm{i}\theta/2} \\ \mathrm{e}^{\mathrm{i}\theta/2} & 0 \end{pmatrix}$$

and rotation of $2\omega t$ in time by

$$\begin{pmatrix} 0 & e^{-i\omega t} \\ e^{i\omega t} & 0 \end{pmatrix} .$$

This matrix operates on a two-level object, called a spinor, e.g. $\{\phi_1 , \phi_2\}$. A spinor system that moves in some general direction, x, say

$$\Phi = \begin{pmatrix} 0 & e^{-i(\omega t - kx)} \\ e^{i(\omega t - kx)} & 0 \end{pmatrix} \begin{pmatrix} \phi_2 \\ \phi_1 \end{pmatrix} \equiv \begin{pmatrix} \phi_1 e^- \\ \phi_2 e^+ \end{pmatrix}$$

(in shorthand notation) is shown to satisfy Schrödinger's [6] and Dirac's [7] equations, by forming the derivatives

$$\frac{\partial \Phi}{\partial t} = i\omega \begin{pmatrix} -\phi_1 e^- \\ \phi_2 e^- \end{pmatrix} , \quad \frac{\partial \Phi}{\partial x} = ik \begin{pmatrix} \phi_1 e^- \\ \phi_2 e^+ \end{pmatrix} , \quad \frac{\partial^2 \Phi}{\partial x^2} = k^2 \begin{pmatrix} -\phi_1 e^- \\ \phi_2 e^+ \end{pmatrix}$$

from which follows that

$$\frac{1}{i\omega} \frac{\partial \Phi}{\partial t} = \frac{1}{k^2} \frac{\partial^2 \Phi}{\partial x^2} .$$

Equating $m = \hbar k^2 / 2\omega$, the equation

$$-i \frac{\partial \Phi}{\partial t} = \frac{\omega}{k^2} \nabla^2 \Phi = \frac{\hbar}{2m} \nabla^2 \Phi \tag{5a}$$

in three dimensions is identical to Schrödinger's equation or its complex conjugate and describes both matter and antimatter, each with the spin states $\{\phi_1, \phi_2\}$. Writing the wave vector $k = 2\pi/\lambda = p/\hbar$, with de Broglie, the quantum condition $\hbar\omega = p^2/2m$ follows directly.

The common assertion that electron spin is a strictly relativistic effect [8] needs modification. As shown before [9] a linearized Schrödinger equation, first order in all space–time derivatives implies the Pauli equation [10] with the correct value of the electron gyromagnetic ratio. The 1/2 spin appears, not as a relativistic effect, but because of the implied four-dimensional space–time. It has the significant advantage that the spin degrees of freedom are contained in the theory from the beginning, without being added ad hoc. However, the spin- and angular-momentum variables remain separated.

Since there is no geometrical understanding of spin in three-dimensional space, it is not reflected in standard wave mechanics. A four-dimensional quaternion function, on the other hand, contains a four-component spinor as temporal characteristic, entangled with the space variables. Spin is generated in four-dimensional quaternion rotation, which is intrinsically different from the axial rotation of three-dimensional space, and is also described as spherical rotation[2] [7]. Because rotation in spin space is governed by one-half the angles of rotation in ordinary space, it has the special property that any entanglement of the spinor with the supporting medium, which develops during rotation of 2π, spontaneously disentangles after a further 2π rotation. This is the result of the half-angle operation of quaternion rotation. It causes a periodic fluctuation in the medium (vacuum) that surrounds the spinor and is observed as spin. The undulation with $\lambda = 2\pi$ on the surface of the unit sphere describes spin as $m\omega = \hbar/2$, half a unit of angular momentum.

Each spin $\begin{pmatrix} e^{i\omega t} \\ 0 \end{pmatrix}$ has an inverse state $\begin{pmatrix} 0 \\ e^{i\omega t} \end{pmatrix}$ and an antispin $\begin{pmatrix} e^{-i\omega t} \\ 0 \end{pmatrix}$. The products

$$(0\ e^{i\omega t}) \begin{pmatrix} e^{i\omega t} \\ 0 \end{pmatrix} = 0, \qquad (e^{-i\omega t}\ 0) \begin{pmatrix} e^{i\omega t} \\ 0 \end{pmatrix} = 1$$

show that spin pairing yields a boson of spin zero, while the annihilation of spin and antispin produces a photon of spin $\hbar$ and which satisfies the wave equation (2a). The condition $c = \sqrt{E/2m}$ $(V = 0)$, now implies conversion of the total mass of matter and antimatter into a photon of energy $E = 2mc^2 = \hbar\omega$.

It is ironic that spin, which is the only non-classical attribute of quantum mechanics, is absent from the pioneering formulations of Heisenberg and Schrödinger. Even in Dirac's equation, the appearance of spin is ascribed by fiat to Lorentz invariance, without further elucidation. In reality, both Lorentz invariance and spin, representing relativity and quantum mechanics, respectively, are properties of the quaternion field that underpins both theories.

4 The Physical Model

Within the time-like domain of Minkowski space, the gravitational field dominates and a quantum-potential field dominates in the space-like domain. Electromagnetic bosons, which occur in the interface, exhibit both time-like and space-like behavior. Events close to, or in, the interface show both quantum and relativistic behavior. There is only one classical–non-classical limit as quantum mechanics and relativity appear as a single theory. Confusion sets in with failure to distinguish between waves and particles as elementary entities and with the possibility of massive objects moving faster than light in the vacuum. It is the unphysical identification of elementary point particles with quantum waves that gives rise to the spurious concepts of wavicle, quantum uncertainty, probability density and infinite self-energy. Elementary wave structures exist without such complications.

[2] The relationship between spherical rotation and quaternions is visualized pictorially by Kauffman[11].

It is instructive to note that the authoritative exposition of Unsöld [12] does not involve probability density.

Standing waves, like gas molecules, exert pressure on the walls of a container. With the container at rest or in uniform motion, the force exerted on any wall is balanced by that exerted on the opposite wall. If the container is accelerated, radiation reflected off the rear wall gains more momentum than that lost by the radiation which reflects off the front wall. It has been demonstrated [13, 14, 16] that radiation exerts a net force, which opposes an applied force on the container, such that

$$\left(m + \frac{E_R}{c^2}\right) a = F .$$

The radiant energy adds an effective inertial mass E_R/c^2 to the mass of the container. In the same way, the total mass of an electron may be interpreted as deriving from internal motion in a phase-locked cavity.

Plane waves with phase velocity v_ϕ form wave packets with group velocity v_g, such that $v_\phi v_g = c^2\ (= 1/\epsilon_0\mu_0)$. The phase and group velocities are the same only for electromagnetic waves with $v_\phi = c$. To describe the inertial properties of a wave packet, a mass variable is introduced [17]. A group velocity $v_g \lesseqgtr c$ defines matter waves as described by the Elbaz [18] equation:

$$\nabla^2 u \pm \left(\frac{m_0 c}{\hbar}\right)^2 u = \frac{1}{c^2}\frac{\partial^2 u}{\partial t^2} \tag{6}$$

This is interpreted to show that the curving of space produces time-like ($v_g < c$) as well as space-like ($v_g > c$) wave packets with real (m) and imaginary (im) mass [19], respectively, known as *bradyons* and *tachyons*.

Equation (6), second order in time, has solutions $u(t)$ and $u(-t)$, defining matter and antimatter waves, respectively. Of the four possible forms, only bradyonic matter waves are directly observed in time-like tangent space. Other types of wave are encountered in subatomic quantum systems.

The formation of wave packets depends on the interaction of a complementary bradyon–tachyon pair. The bradyonic group velocity corresponds to the de Broglie wavelength of the packet, $\lambda_{dB} = h/mv_g$. The tachyonic component defines the internal structure of the wave packet with Compton wavelength $\lambda_C = h/mc$. The two components are said [20] to be trapped in a relativistically invariant way. We note that $v_b v_t = c^2 = 1/\epsilon_0\mu_0$, where the group velocity of the tachyon v_t matches the phase velocity of the bradyon. Such a wave packet is not dispersive. Essentially the same model, developed from another perspective, is described by Wolff [21–23]. An elementary unit of charge (e.g. electron) is considered here as a standing wave packet created by the interaction between a pair of time-symmetric spherical scalar waves: a retarded wave radiated by the electron, in balance with an advanced wave that represents the resultant as the radiation from all other sources in the cosmos interferes. This is Mach's principle in wave formalism.

Variability in the internal structure of elementary wave packets is responsible for the appearance of protons, neutrons and electrons [24], which interact by the exchange of intermediate massless photons. Chemical interactions, all of this type [6], are responsible for the growth of massive structures, which on the macroscale interact gravitationally.

All debates on the interpretation of quantum mechanics must end in confusion, unless the classical and non-classical models of the world are clearly distinguished. The classical model is based on the assumption that persistent fragmentation of matter terminates in a set of elementary particles that resist further subdivision, but retain the innate quality to predict the behavior of matter in the bulk. A non-classical alternative starts at the other extreme with a featureless plenum that develops periodic wave structures in a topologically closed universe. In projective relativity [25], there is

> ... no such thing as a body in space, but matter is an aspect of the space–time structure.

These elementary waves coalesce into bigger units that exhibit all the known properties of ponderable matter.

Classical mechanics analyses the interaction between particles, and non-classical mechanics should study the interaction between wave structures. We repeat that the two models do not refer to classical and non-classical domains—they both model the same world, but from different points of view. It so happens that at different levels of aggregation, one or the other provides a more convenient description. Attempts to describe classical structures non-classically, or vice versa, inevitably end up with illogical conclusions.

Physics has the dilemma of irrefutable evidence for a four-dimensional world, but a genetic inability among physicists to visualize more than three dimensions. It is therefore not surprising to find that those instances, in which reality is badly distorted in three-dimensional projection, inevitably lead to convoluted theories, bordering on the supernatural. Quantum mechanics is a prime example of such a theory. It was inspired by experimental results that defied explanation based on classical theory. It was first recognized in the study of microphysical systems, which in time came to be seen as deviating from the classical and therefore subject to a new theory, without relevance in macrophysics.

A more plausible interpretation is that the motion of ponderous objects, projected into tangent three-dimensional space, differs imperceptibly from four-dimensional reality in the local environment where a classical description suffices. It only becomes an issue for fast-moving objects and where particle mass approaches zero. The real meaning of both relativity and quantum theory is obscured by their formulation as alternatives to Newtonian mechanics that kick in at some classical limit.

A notable difference between three- and four-dimensional formulation of quantum mechanics occurs in their commutation properties. In $\mathbb{R}^3$ commutators that involve a time-dependent variable are found to be non-zero. For instance,

$$[p,q] = (pq - qp) = \mathrm{i}\hbar, \quad [J_x, J_y] = \mathrm{i}\hbar J_z, \text{ etc.}$$

Uncertainty relationships such as $\Delta p \Delta q > \hbar$, $\Delta E \Delta t > \hbar$, etc., derive directly from these commutation rules [26].

The unexpected appearance of complex operators is also associated with non-zero commutators and reflects the essential two-dimensional representation in $\mathbb{M}^2$ Minkowski space–time. In four-dimensional space–time, $\mathbb{H}^4$, all commutators are non-zero, as appropriate for wave motion of both quantum and relativity theories. An important consequence is that local observation has no validity on global extrapolation, as evidenced by the appearance of cosmical red shifts in the curved manifold and the illusion of an expanding universe.

5 The Chemical Model

Chemical theory, if anything, is distorted even more than physics on projection from four-dimensional space–time. In electromagnetic and other field theories, gauge particles have mathematically assigned phase factors, which in chemistry are simulated as probability density. Whereas the purely mathematical symbolism suffices as working models in particle physics, chemistry has the more stringent demand to deal with extended three-dimensional entities. Even at its lowest level, the known chemical function of an electron, defined as a structureless point particle, becomes incomprehensible.

The analysis of dynamic systems in terms of point particles originated with Newton, but the context in which the concept was introduced has been ignored and forgotten. Newton was concerned with the motion of heavenly bodies and their mutual interaction. To avoid the complication that the moon's effect at the proximal and distal surfaces of the earth cannot be the same, the total interaction was assumed mathematically equivalent to the attraction between objects with all mass concentrated at their respective centers of gravity, which is a point. Earth and moon are therefore modeled as interacting mass points. What works for planets also works for apples and electrons. In practical application, reference to the center of gravity was made less frequently, and, in time, the actual nature of elementary units of matter as mass points became generally accepted as physical reality.

When the wave nature of electrons was discovered experimentally, Newton's perception of mass points as centers of gravity had been inactive for so long that, instead of a wave packet with a center of gravity, an electron was considered as the physical union of a particle and a wave. In this approach, particle nature is considered to be of primary importance, and wave nature is simulated by a probability function that specifies particle coordinates. This interpretation was pioneered by de Broglie's postulate [27] of a particle piloted by a wave. However, the more natural wave model does the opposite and follows the motion of a wave packet in terms of its center-of-gravity coordinates. The photoelectric and Compton effects can then be understood as interaction between waves, rather than particles, as pointed out by Schrödinger [28]. As stated in the abstract of [28]:

> A definite ψ-distribution in configuration space is interpreted as a continuous distribution of electricity (and electric current density) in actual space.

Unfortunately, this level-headed interpretation was never generally accepted. As remarked by Bohm [29]:

> The current formulation of quantum mechanics must be regarded merely as a statistical algorithm, which provides no conceptual structure in terms of which the movement of individual systems can be understood.

It is only the theory of general relativity that provides any guidance towards an understanding of the elementary nature of matter, via Einstein's field equations, conveniently condensed into the form

$$G_{\mu\nu} = kT_{\mu\nu}, \quad \mu, \nu = 0, 3\,,$$

which balances the tensor of space–time curvature against the stress tensor of the matter–energy field. The vital assertion is that a vanishing curvature tensor, which implies flat Euclidean space, demands the disappearance of all matter. It is unequivocally inferred that matter appears as a result of the curving of space–time. By the minimal assumption, known as Occam's razor, matter must be identified as a distortion of curved space–time.

As a reasonable conjecture, we now propose that curved space–time, like an inflexible sheet wrapped around a curved surface, must develop persistent wrinkles—the elementary units of matter or energy. We envisage flat space–time in featureless undulation that develops elementary wave packets when curved. We recognize few types of wave packet with internal wave patterns perceived as the characteristic mass, charge, spin and chirality of the four-dimensional elementary units whose behavior is prescribed by a potential function according to Eq. (2).

On projection into three-dimensional tangent space, the chiral forms known as matter and antimatter are distinct and, as $c \to \infty$ [30], obey conjugate forms of the wave equation (2a), the three-dimensional approximation of (2). To substantiate this reasoning, it is noted that many features of the H atomic spectrum are reproduced to good approximation by the solution of (5) for the single electron on a hydrogen atom.

The crunch comes when trying to analyse non-classical many-electron systems by the same procedure. The mathematics to solve the many-body differential equation does not exist. The popular alternative is to consider each electron as an individual particle and to describe an n-electron system by a probability density in $3n$-dimensional configuration space. The use of complex variables is tacitly avoided. The result is a procedure that pretends to simulate a non-classical problem by a classical model, with an unnecessary complicated structure, designed to resemble quantum formalism. In this case, the statistical model that works for an ideal gas fails to explain the behavior of a many-electron wave.

An alternative procedure is suggested by the recognition of elementary matter as wave-like distortions of space–time. Unlike free-floating hard particles in a void, the wave packets envisaged here remain part of the medium, and their distribution therefore depends on the symmetry of space–time. The extranuclear charge cloud on an atom may thus be viewed as the coalescence of electronic wave packets to constitute a common spherical standing wave around the nucleus. The internal structure of the wave must reflect the charge distribution as optimized under the nuclear attraction, like a three-dimensional analogue of the essentially planar solar system.

Optimization by a golden spiral predicts the correct distribution of matter in the solar system [31], with the inference that the spiral structure reflects space–time topology. Fractal models of the universe, which imply cosmic self-similarity, would then indicate the same optimization for extranuclear electron density. The resulting wave structure inevitably carries an imprint of the golden ratio.

Interatomic interaction entails the interference of extranuclear electronic waves. Constructive interference must occur at specific interatomic distances, which should correlate with the notion of bond order, numerically related to the golden ratio. The feasibility of modeling chemical interaction by elementary number theory is foreseen.

6 Conclusion

Contrary to popular belief, new ideas in science are rarely embraced with acclaim. This was known, in a different context, to Machiavelli who stated 500 years ago [32]:

> … that there is nothing more difficult to arrange, more doubtful of success, and more dangerous to carry through than initiating changes …. The innovator makes enemies of all those who prospered under the old order, and only lukewarm support is forthcoming from those who would prosper under the new. Their support is lukewarm partly from fear of their adversaries, who have the existing laws on their side, and partly because men are generally incredulous, never really trusting new things unless they have tested them by experience. In consequence, whenever those who oppose the changes can do so, they attack vigorously, and the defence made by the others is only lukewarm. So both the innovator and his friends are endangered together.

As an example in science, Schrödinger had to endure such attacks in response to his wave-mechanical interpretation of quantum effects, as evidenced by his statement, from the biography by Moore [33]:

> Let me say at the outset, that in this discourse, I am opposing not a few special statements of quantum mechanics held today (1950), I am opposing as it were the whole of it, I am opposing its basic views that have been shaped 25 years ago, when Max Born put forwards his probability interpretation, which was accepted by almost everybody. … I don't like it, and I'm sorry I ever had anything to do with it.

Schrödinger tried to introduce a wave model of matter in opposition to the contemporary dogmatic belief in elementary particles. He failed and the debate was closed for the best part of a century. By now, the ranks of those who profit by the preservation of the wave-particle model have swelled by orders of magnitude. Even the few lukewarm defenders have to rely on commercialized software based on probabilities to generate results considered suitable for publication in the mainstream media. There is no reason to believe that the 4D analysis proposed here will be treated any differently. Promoting a new system remains fraught with danger, despite compelling mathematical support.

Current consensus among mathematicians [34] is that normed division algebras, which could be the basis of high-dimensional physical vector spaces, are restricted to one, two, four and eight dimensions. We interpret this to explain why physical theories in three dimensions are plagued by confusing features such as non-commutation and complex phases that intrude themselves in quantum theory. In four dimensions, these would be natural features. The same argument explains why five-dimensional Kaluza–Klein models fail as unified field theories. The alternative four-dimensional unification based on projective geometry in curved space–time [35] and a wave model of matter works without awkward compacted dimensions and develops gauge invariance in a natural way.

Some common practices further aggravate the situation. The accepted interpretation of special relativity considers all space outside of the Minkowski time cone as non-physical. This prejudice obscures the non-local nature of quantum theory and distorts the common perception of space–time topology. By an equally arbitrary assumption, advanced solutions (in $-t$) of the three-dimensional wave equation are rejected. This way all perceptions of space–time chirality, the existence of antimatter and non-local correlation are lost.

In chemical theory, misreading of the superposition principle underpins the widespread use of real orbitals and basis sets, without any mathematical meaning. Half a century's research results in quantum chemistry may well be wasted effort. But this represents Machiavelli's profit under the old system. We propose that the utility of number theory in the description of chemical systems could provide an escape route from this dilemma.

Acknowledgements I have discussed the mathematics of four dimensions many times with my colleague Casper Schutte, and I gratefully acknowledge his valuable input.

Appendix A: Hypercomplex Numbers

The theory of special relativity is conveniently summarized by a set of equations, known as a Lorentz transformation, which describes all relative motion, including that of electromagnetic signals, observed to propagate with constant speed c, irrespective of the observer's state of motion. This transformation,

$$x' = \frac{x - vt}{\sqrt{1 - v^2/c^2}}, \qquad t' = \frac{t - vx/c^2}{\sqrt{1 - v^2/c^2}},$$

relates two frames of reference in relative motion and has no meaning in a system that separates space and time variables. The resulting four-dimensional space–time is known as Minkowski space.

From the equation for the moving front of a spherical light wave

$$x^2 + y^2 + z^2 = c^2t^2 \,,$$

a fourth coordinate is often defined as $x_0 = ct$, redefined by Minkowski as $x_0 = \mathrm{i}ct$, such that

$$x_0^2 + x_1^2 + x_2^2 + x_3^2 = 0$$

at the wavefront. In terms of the velocity ratio $\beta = v/c$ and $\gamma = 1/\sqrt{1-\beta^2}$, the Lorentz transformation for uniform relative motion along x_1 then takes the simple form, $\mathbf{x}' = \mathbf{Lx}$, i.e.

$$\begin{pmatrix} x_0' \\ x_1' \end{pmatrix} = \begin{pmatrix} \gamma & -\mathrm{i}\beta\gamma \\ \mathrm{i}\beta\gamma & \gamma \end{pmatrix} \begin{pmatrix} x_0 \\ x_1 \end{pmatrix}$$

This transformation matrix has the same form as an orthogonal rotation matrix

$$\mathbf{R} = \begin{pmatrix} \cos\phi & -\sin\phi \\ \sin\phi & \cos\phi \end{pmatrix} .$$

The Lorentz transformation thereby defines a rotation in the (x, t) plane through an imaginary angle ϕ, defined by

$$\cos\phi = \frac{1}{\sqrt{1-\beta^2}}, \qquad \sin\phi = \frac{\mathrm{i}\beta}{\sqrt{1-\beta^2}}, \qquad \phi = \tan^{-1}\mathrm{i}\beta \,.$$

As this rotation mathematically interchanges time and space coordinates, it means that they are symmetry related and no longer separable in the usual way. It is therefore more appropriate to deal with four-dimensional space–time, rather than the traditional three-dimensional space and absolute time. To visualize Minkowski space, it is useful first to review some properties of the complex plane.

Appendix B: Complex Numbers

There is a similarity between two-dimensional vectors and complex numbers, but also subtle differences. One striking difference is between the product functions of complex numbers and vectors.

The product of two vectors is either a scalar

$$z_3 = \mathbf{z_1} \cdot \mathbf{z_2} = z_1 z_2 \cos\theta = x_1 x_2 + y_1 y_2$$

or a vector

$$\mathbf{z}_3 = \mathbf{z}_1 \times \mathbf{z}_2 = z_1 z_2 \sin\theta = y_1 x_2 - y_2 x_1 \,.$$

By way of comparison, the product of two complex numbers is

$$\begin{aligned} z_1 z_2 &= (x_1 + \mathrm{i}y_1)(x_2 + \mathrm{i}y_2) \\ &= (x_1 x_2 - y_1 y_2) + \mathrm{i}(x_1 y_2 + x_2 y_1) \\ z_3 &= x_3 + \mathrm{i}y_3 \,. \end{aligned}$$

The complex product contains two terms, not unlike the scalar and vector products, from which it differs only in a sign convention.

It is well known that the complex number $x+\mathrm{i}y$ is given in polar form by Euler's equation:

$$r(\cos\theta + \mathrm{i}\sin\theta) = r\mathrm{e}^{\mathrm{i}\theta} \,.$$

The product

$$z_1 z_2 = r_1 \mathrm{e}^{\mathrm{i}\theta_1} \cdot r_2 \mathrm{e}^{\mathrm{i}\theta_2} = r_1 r_2 \mathrm{e}^{\mathrm{i}(\theta_1+\theta_2)}$$

is recognized immediately as the rotation of z_1 through the angle θ_2 and increase of its length by a factor r_2. To summarize, an operator $w = u + \mathrm{i}v = C\mathrm{e}^{\mathrm{i}\theta}$, with $C^2 = u^2 + v^2$, $\theta = \tan^{-1}(v/u)$, when it multiplies any vector, increases its length by a factor C. The magnitude and phase of the complex number $x + \mathrm{i}y$ are $r = \sqrt{x^2 + y^2}$ and $\theta = \tan^{-1}(y/x)$. The complex conjugate $z^* = (x + \mathrm{i}y)^* = x - \mathrm{i}y$ has the phase $-\theta$ and magnitude r. The magnitude of a complex quantity is obtained from $zz^* = (r\mathrm{e}^{\mathrm{i}\theta})(r\mathrm{e}^{-\mathrm{i}\theta}) = r^2$, which is always real and nonnegative.

Matrices in the form $\begin{pmatrix} \alpha & -\beta \\ \beta & \alpha \end{pmatrix}$, combined by matrix addition and multiplication, are isomorphic with the field of complex numbers $(\alpha + \mathrm{i}\beta)$. This way,

$$(\alpha + \mathrm{i}\beta) + (\gamma + \mathrm{i}\delta) = (\alpha + \gamma) + \mathrm{i}(\beta + \delta) \,,$$

and

$$\begin{pmatrix} \alpha & -\beta \\ \beta & \alpha \end{pmatrix} + \begin{pmatrix} \gamma & -\delta \\ \delta & \gamma \end{pmatrix} = \begin{pmatrix} \alpha + \gamma & -(\beta + \delta) \\ \beta + \delta & \alpha + \gamma \end{pmatrix} .$$

Also

$$(\alpha + \mathrm{i}\beta)(\gamma + \mathrm{i}\delta) = (\alpha\gamma - \beta\delta) + \mathrm{i}(\beta\gamma + \alpha\delta)$$

and

$$\begin{pmatrix} \alpha & -\beta \\ \beta & \alpha \end{pmatrix} \begin{pmatrix} \gamma & -\delta \\ \delta & \gamma \end{pmatrix} = \begin{pmatrix} \alpha\gamma - \beta\delta & -(\beta\gamma + \alpha\delta) \\ \beta\gamma + \alpha\delta & \alpha\gamma - \beta\delta \end{pmatrix} .$$

Euler's equation in the form

$$r\mathrm{e}^{\mathrm{i}\theta} = r \begin{pmatrix} \cos\theta & -\sin\theta \\ \sin\theta & \cos\theta \end{pmatrix}$$

confirms the geometrical meaning of complex numbers as rotation and enlargement. As an example, the complex number i corresponds to the matrix $\begin{pmatrix} 0 & -1 \\ 1 & 0 \end{pmatrix}$, which represents a counterclockwise rotation of $\pi/2$ about the origin.

Another alternative form of the rotation matrix is obtained by setting $\alpha = a^2 + b^2$ and $\beta = 2ab$, with $a = \cos(\theta/2), b = \sin(\theta/2)$. Hence,

$$\alpha = \cos^2(\theta/2) - \sin^2(\theta/2) = \cos\theta$$

$$\beta = 2\cos(\theta/2)\sin(\theta/2) = \sin\theta\,,$$

i.e.

$$(a,b) = \begin{pmatrix} a^2 - b^2 & -2ab \\ 2ab & a^2 - b^2 \end{pmatrix} = \begin{pmatrix} \cos\theta & -\sin\theta \\ \sin\theta & \cos\theta \end{pmatrix}. \tag{A.1}$$

Appendix C: Quaternions

Extension of the complex formalism to more dimensions suggests the definition of related *hypercomplex* numbers. On multiplication of two three-dimensional vectors, without defining the mathematical properties of unit vectors **i**, **j**, **k**, the formal result is

$$\begin{aligned} q &= (\mathbf{i}x_1 + \mathbf{j}y_1 + \mathbf{k}z_1)(\mathbf{i}x_2 + \mathbf{j}y_2 + \mathbf{k}z_2) \\ &= \mathbf{i}^2 x_1 x_2 + \mathbf{j}^2 y_1 y_2 + \mathbf{k}^2 z_1 z_2 \\ &\quad + \mathbf{ij}x_1 y_2 + \mathbf{ji}y_1 x_2 + \mathbf{ik}x_1 z_2 + \mathbf{ki}z_1 x_2 + \mathbf{jk}y_1 z_2 + \mathbf{kj}z_1 y_2 \end{aligned}$$

This expression is rearranged into the same form as a complex product by defining

$$\mathbf{i}^2 = \mathbf{j}^2 = \mathbf{k}^2 = -1$$

$$\mathbf{ij} = \mathbf{k}\,,\ \mathbf{jk} = \mathbf{i}\,,\ \mathbf{ki} = \mathbf{j}$$

$$\mathbf{ji} = -\mathbf{k}\,,\ \mathbf{kj} = -\mathbf{i}\,,\ \mathbf{ik} = -\mathbf{j}\,,$$

the result first obtained by William Hamilton who defined

$$q = -(x_1x_2 + y_1y_2 + z_1z_2) + \mathrm{i}(y_1z_2 - y_2z_1) + \mathrm{j}(z_1x_2 - z_2x_1) + \mathrm{k}(x_1y_2 - x_2y_1)$$

with the rule of composition: $\mathrm{i}^2 = \mathrm{j}^2 = \mathrm{k}^2 = \mathrm{ijk} = -1$.
A hypercomplex number of unit norm can now be defined in the form $z = a_0 + a_i e_i$, where the e_i are generalizations of $\sqrt{-1}$, in matrix notation:

$$1 = \begin{pmatrix} 1 & 0 \\ 0 & 1 \end{pmatrix}, \quad e_1 = \begin{pmatrix} \mathrm{i} & 0 \\ 0 & -\mathrm{i} \end{pmatrix}, \quad e_2 = \begin{pmatrix} 0 & 1 \\ -1 & 0 \end{pmatrix}, \quad e_3 = \begin{pmatrix} 0 & \mathrm{i} \\ \mathrm{i} & 0 \end{pmatrix}$$

For i = 1, 3, it is called a quaternion and the matrices are quaternion units. The unit number **1** represents the unit vector in the fourth dimension.

To represent a quaternion in matrix form, as was done for complex numbers, it is written as a pair of complex numbers, $q = (u, v)$, in the same way that the complex number $a + ib$ is written as $(a, b) = \begin{pmatrix} a & -b \\ b & a \end{pmatrix}$. As for complex numbers, the product of (u, v) and its conjugate should be $(|u|^2 + |v|^2, 0)$. This result is obtained by defining the conjugate $q^* = (u^*, -v)$ and the matrix equivalent of (u, v) as $\begin{pmatrix} u & v^* \\ -v & u^* \end{pmatrix}$. This way,

$$qq^* = \begin{pmatrix} u & v^* \\ -v & u^* \end{pmatrix} \begin{pmatrix} u^* & -v \\ v & u \end{pmatrix} = \begin{pmatrix} uu^* + v^*v & -uv^* + v^*u \\ -vu^* + u^*v & vv^* + u^*u \end{pmatrix}$$
$$\equiv \left(|u|^2 + |v|^2, 0\right)$$

The modulus of q, $\sqrt{q^*q} = \sqrt{u^2 + v^2}$. Notice that the determinant of the matrix of $q = (u, v)$ gives

$$\begin{vmatrix} u & v^* \\ -v & u^* \end{vmatrix} = |u|^2 + |v|^2 .$$

By expanding the quaternion in terms of the complex pair, $u = a + ib$ and $v = c + id$, it follows directly that $q^2 = a^2 + b^2 + c^2 + d^2$. In terms of four quaternion components,

$$q \cdot q = q_0^2 + q_1^2 + q_2^2 + q_3^2 . \tag{A.2}$$

For $|u|^2 + |v|^2 = 1$, the product of the quaternion and its conjugate (or adjoint) is the unitary matrix $\begin{pmatrix} 1 & 0 \\ 0 & 1 \end{pmatrix}$. Unit quaternions (of modulus 1) are therefore seen to define rotations of $\mathbb{C}^2$, just as the unit complex numbers correspond to rotations of $\mathbb{R}^2$. The group of unitary transformations of $\mathbb{C}^2$ of determinant 1 is the special unitary Lie group $SU(2)$, which describes the spin function.

In Hamilton's notation, the conjugate of the general quaternion $q = a + \mathrm{i}b + \mathrm{j}c + \mathrm{k}d$ is $q^* = a - \mathrm{i}b - \mathrm{j}c - \mathrm{k}d$, so that the square of the magnitude $|q|^2 = q^*q = a^2 + b^2 + c^2 + d^2$ is a simple extension of the rule for complex numbers.

As with complex numbers, the product of two quaternions

$$(p = \alpha + \mathrm{i}\beta + \mathrm{j}\gamma + \mathrm{k}\delta)q = (\alpha a - \beta b - \gamma c - \delta d) + \mathrm{i}(\alpha b + \beta a + \gamma d - \delta c)$$
$$+\mathrm{j}(\alpha c - \beta d + \gamma a + \delta b) + \mathrm{k}(\alpha d + \beta c - \gamma b + \delta a)$$

is another quaternion. Quaternion multiplication is clearly not commutative, as $pq \neq qp$.

By expanding the exponential and using the rules for multiplication of i, j, k, it can be shown that, for $\alpha^2+\beta^2+\gamma^2 = 1$, a quaternion equivalent of Euler's equation for the imaginary exponent is obtained as

$$e^{\theta(i\alpha+j\beta+k\gamma)} = \cos\theta + \sin\theta(i\alpha + j\beta + k\gamma)\,.$$

This exponential is of unit length; the exponent represents a space vector of length θ, with direction as given by the direction cosines $\alpha,\ \beta,\ \gamma$.

Any quaternion can now be represented in the form $Qe^{\theta(i\alpha+j\beta+k\gamma)}$, where Q is the length of a four vector with direction as fixed by $\theta,\ \alpha,\ \beta,\ \gamma$. By analogy with the complex exponential, it is anticipated that this operator (with $Q = 1$) describes a rotation. The argument on which it operates is also a quaternion. In three dimensions, the vector $f = ix + jk + kz$, in quaternion notation, is rotated by an angle θ about an axis of direction cosines $\alpha,\ \beta,\ \gamma$ into f' according to

$$f' = ix' + jy' + kz' = e^{(\theta/2)(i\alpha+j\beta+k\gamma)} f e^{-(\theta/2)(i\alpha+j\beta+k\gamma)} = q\cdot f\cdot q^{-1}\,,$$

where $\alpha^2 + \beta^2 + \gamma^2 = 1$.

As in (A.1), we define the rotation matrix by a pair of complex numbers (u, v), setting

$$\begin{aligned} q_0 &= a\cos(\theta/2)\\ q_1 &= i\sin(\theta/2)\\ q_2 &= j\sin(\theta/2)\\ q_3 &= k\sin(\theta/2)\,, \end{aligned}$$

which define the three-dimensional rotation matrix:

$$\mathbf{R}(q) = \begin{pmatrix} q_0^2+q_1^2-q_2^2-q_3^2 & 2q_1q_2-2q_0q_3 & 2q_1q_3+2q_0q_2\\ 2q_1q_2+2q_0q_3 & q_0^2-q_1^2+q_2^2-q_3^2 & 2q_2q_3-2q_0q_1\\ 2q_1q_3-2q_0q_2 & 2q_2q_3+2q_0q_1 & q_0^2-q_1^2-q_2^2+q_3^2 \end{pmatrix} \tag{A.3}$$

Each row (or column) has unit magnitude, e.g.:

$$\left(q_0^2+q_1^2-q_2^2-q_3^2\right)^2 + (2q_1q_2-2q_0q_3)^2 + (2q_1q_3+2q_0q_2)^2 = (q\cdot q)^2 = 1\,,$$

and the total matrix $\left(\mathbf{R}f = qfq^{-1}\right)$ is orthogonal, as required for 3D rotation.

Quaternions describe rotation in any number of dimensions from 1 to 4. It is straightforward to demonstrate[3] that for the special case of rotation about the x-axis,

[3]Using $\sin(\theta/2) = \pm\sqrt{\frac{1}{2}(1-\cos\theta)}$, $\cos(\theta/2) = \pm\sqrt{\frac{1}{2}(1+\cos\theta)}$.

$$
\begin{aligned}
f' &= e^{(\theta/2)i}(ix + jy + kz)e^{-(\theta/2)i} \\
&= ix + j(y\cos\theta - z\sin\theta) + k(y\sin\theta + z\cos\theta) \\
&= \begin{matrix}(i\ j\ k)\end{matrix} \begin{pmatrix} 1 & 0 & 0 \\ 0 & \cos\theta & -\sin\theta \\ 0 & \sin\theta & \cos\theta \end{pmatrix} \begin{pmatrix} x \\ y \\ z \end{pmatrix}
\end{aligned}
$$

i.e. a rotation through the imaginary angle θ about the x-axis in the (y,z)-plane. The right-to-left order of matrix multiplication is important. Proof of the general case only involves more algebra. Symbolically $f' = \zeta f \zeta^*$. By a second rotation, $f'' = \eta f' \eta^* = \eta\zeta f \zeta^* \eta^*$.

The final result depends on the order in which the operations are applied, because of the fact that the quaternions η and ζ do not commute. The quantity Q is called the *tensor* (stretcher), and the exponential is called the *versor* (turner) of the operator.

In four-dimensional rotation, the argument of the operation is the full quaternion four-vector, $v_\mu = (v_0, v_i)$, rather than the three-vector f with $v_0 = 0$, considered before. On working out the full rotation matrix, it turns out to decompose into a pair of 3D rotations, such as (A.3), indicating that the four-dimensional rotation amounts to double covering of the underlying space of 3D rotations [37].

References

1. Dirac PAM (1928) Proc Roy Soc A 117:610
2. Taub AH, Veblen O, von Neumann J (1934) Proc Nat Acad Sci USA 20:383
3. Dirac PAM (1938) Proc Roy Soc A167:148
4. Morse PM, Feshbach H (1956) Methods of theoretical physics. McGraw-Hill, New York
5. Misner CW, Thorne K, Wheeler JA (1973) Gravitation. Freeman, San Francisco, p 1143
6. Boeyens JCA (2008) Chemistry from first principles. Springer, Dordrecht, Heidelberg, London, New York
7. Batty-Pratt EP, Racey TJ (1980) Int J Theor Phys 19:437
8. Carmeli M (1985) Nuovo Cimento Lett 42:67
9. Lévy-Leblond J-M (1967) Comm Math Phys 6:286
10. Pauli W (1927) Z Phys 43:601
11. Kauffman LH (1987) On knots. Princeton University Press, Princeton
12. Unsöld A (1927) Ann. Phys 82:355
13. Bohm D (1996) The special theory of relativity. Routledge, London
14. Einstein A (1911) On the influence of gravitation on the propagation of light. Translation of German original. In: Perrett P, Jeffery GB (translators) (1952) The principle of relativity. Dover, New York
15. Perrett P, Jeffery GB (translators) (1952) The principle of relativity. Dover, New York
16. Jennison RC, Drinkwater AJ (1977) J Phys A 10:167
17. Elbaz C (1986) Phys Lett 114A:445
18. Elbaz C (1983) CR Acad Sci Paris 297:455
19. Feinberg G (1967) Phys Rev 159:1089
20. Corben HC (1977) Lett Nuovo Cimento 20:645
21. Wolff M (1993) Phys Essays 6:181
22. Wolff M (1995) Galilean Electrodyn 6:83
23. Wolff M(1997) Front Perspect 4:67

24. Jennison RC (1978) J Phys A Math Gen 11:1525
25. McCrea WH (1937) Supplement to Nature of June 12, 139:1002 (Quoting Veblen and Whitehead)
26. Bohm D (1951) Quantum theory. Dover reprint, Prentice-Hall, Upper Saddle River
27. de Broglie L (1926) Nature 118:441
28. Schrödinger E (1978) Collected papers on wave mechanics. Translated from the 1927 German edition, Chelsea, NY, p 124
29. Bohm D, Bub J (1966) Revs Mod Phys 38:453
30. Oppenheimer JR (1930) Phys Rev 35:461
31. Boeyens JCA (2009) Phys Essays 22:493
32. Machiavelli N (1975) The prince. Penguin Books, Hammondsworth. Translated from the 1929 Italian edition by George Bull
33. Moore W (1989) Schrödinger—life and thought. University Press, Cambridge
34. Gowers T (ed) (2008) The Princeton companion to mathematics. University Press, Princeton
35. Veblen O (1933) Projektive Relativitätstheorie. Springer, Berlin. English translation in Boeyens JCA (2010) Chemical cosmology. Springer.com
36. Boeyens JCA (2010) Chemical cosmology. Springer, Dordrecht, Heidelberg, London, New York
37. Hanson AJ (2006) Visualizing quaternions. Elsevier (Morgan Kaufman), Amsterdam

Struct Bond (2013) 148: 49–69
DOI: 10.1007/978-3-642-31977-8_3
© Springer-Verlag Berlin Heidelberg 2013

Is the Rydberg–Ritz Relationship Valid?

C.J.H. Schutte

Abstract In the first part of this chapter, it is shown that the linear relationship between the energy E_n of any quantum state of the hydrogen atom and the *negative inverse square of the quantum number n* can be used, together with the Rydberg–Ritz combination principle, to provide an internal check of its own validity, utilizing the most accurate atomic spectral data. This internal check uses the fact that the value of the linear proportionality constant can be obtained both from the *slope* and from the *intercept* of the straight line on the energy axis. If these two values differ by more than that allowed by experimental scatter, there is serious doubt about the validity of the inverse-squared relationship. This analysis shows that the relationship is nearly but not exactly satisfied. In the second part of this chapter, it is shown that the usual interpretation of the inverse-squared relationship obscures the fact that it actually leads to imaginary values for the quantum numbers and not to the real integral values as assumed up to now. Both analyses indicate that *nonrelativistic quantum mechanics* is not based upon solid foundations as assumed up till now, and requires critical reexamination, especially with respect to the role of time.

Keywords Atomic spectroscopy · Balmer formula · Hydrogen spectrum · Quantum theory · Rydberg constant

Contents

C.J.H. Schutte (✉)
Unit for Advanced Scholarship and Department of Chemistry, University of Pretoria, Lynnwood Road, 0002 Pretoria, South Africa
e-mail: casper.schutte@up.ac.za

1 Introduction

A linear relation, which I call the *fundamental relation of atomic spectroscopy*, originating from the empirical atomic spectroscopy of the free hydrogen atom, lies at the heart of *nonrelativistic quantum theory*. This relation, which neglects the hyperfine interactions in the H-atom, which leads to the so-called hyperfine splitting of the individual atomic spectral lines, can be stated as follows:

> The *fundamental relation of atomic spectroscopy* is the statement that the energy of any electronic state of the *hydrogen atom*, E_n, is proportional to the negative inverse of the square of an *integral* number, n, where $n = \{1, 2, 3, \cdots, \infty\}$.

It does not matter in which unit the energy states E_n of the hydrogen-atom lines are measured, but it is customary to display the energy in terms of joule J, wave number $\overline{\nu}$, or wavelength λ (interconversion factors are $1/hc$ to go from J to wave number, and $\overline{\nu} = 1/\lambda$ for wavelength to wave number). In this chapter, I use a new convention, namely, to put a bar over all energies, which are measured in wave numbers, $\overline{\nu}$; in particular, this applies to any constants of proportionality that appear in energy relationships.

It is the ideal of *nonrelativistic quantum theory* to derive the fundamental relation of atomic spectroscopy from a physical model, underpinned by a set of (up to now, incomplete) assumptions. The following question needs to be asked:

> Is the fundamental relation of atomic spectroscopy inherently able to confirm the validity of the assumptions and thus that of the derived energy equations of the nonrelativistic hydrogen-atom models of Bohr, Schrödinger, and Heisenberg?

It is thus the object of this chapter to determine whether this fundamental relation of quantum theory is indeed a *suitable ideal equation* to use to test the "validity" or "suitability" of theoretically derived equations and their respective underpinning models and assumptions.[1] Two different methods are employed here to probe the

[1] It is not the aim of this chapter to repeat any of the theoretical derivations of this equation nor to give a critical assessment of the model and its assumptions and mathematical intricacies here, since it is repeated in every textbook dealing with the subject.

suitability of this fundamental equation, namely, *firstly*, an analysis based upon recent accurate spectroscopic data and the internal structure of the Rydberg–Ritz combination principle and, *secondly*, an analysis of the mathematical consequences of the fundamental equation. Both these methods are new, as far as can be determined.

This *integral number n* of the fundamental relation is identified with the *main quantum number* in all nonrelativistic theoretical models of the hydrogen atom, and it is included in the set of four *theoretical quantum numbers* $\{n, \ell, m_\ell, m_s\}$ that supposedly uniquely describes the energy states of its single electron. This fundamental relation, written in energy units, is given by the expression

$$E_n = -R\frac{1}{n^2}, \quad n = \{1, 2, 3, \ldots, \infty\}, \tag{1}$$

where R is a universal constant of proportionality, commonly known as the Rydberg constant, the ideal value of which is given by CODATA and IUPAC (see [1–3]) in terms of wave numbers as

$$\overline{R}_\infty = 109,737.31568527(73)\,\mathrm{cm}^{-1} \tag{2}$$

and the origin of the infinity sign is explained below. $\overline{R}_\infty$ is considered to be the best-determined physical constant of all physical constants.

The theoretical calculation of the Rydberg constant starts out by assuming the correctness of Eq. (1) for the field-free hydrogen atom. Since this expression is identical with the expressions derived by Bohr in 1915, it is warranted to accept the values as deduced by him. This means inter alia the acceptance of the constant called the radius of the first Bohr orbit, a_0, which is given by

$$a_0 = (4\pi\epsilon_0)\cdot\frac{h^2}{4\pi^2 m_\mathrm{e}\mathrm{e}^2} = 5.2917720859(36)\times 10^{-11}\,\mathrm{m}.$$

It is important to note that the electron mass m_e occurs here as if it were the only mass in the system electron–proton, that is, it is assumed that the electron is bound to an *infinite mass* called the nucleus, which is naturally not true; the *reduced mass* of the system electron–proton is introduced below to take this into account. This infinite mass is the origin of the infinity symbol occurring in the Rydberg constant, $\overline{R}_\infty$.

For the fulfillment of Eq. (1), we need to find the theoretically calculated energy of the hydrogen atom in its electronic ground state. This is given by the Bohr theory as E_H, which is found to be

$$E_\mathrm{H} = \frac{h^2}{(4\pi^2)\, m_\mathrm{e}\alpha_0^2} = 4.35974394(22)\times 10^{-18}\,\mathrm{J}\,.$$

It is clear that the accuracy with which the fundamental constants h, m_e, the electron charge e, and a_0 are known also determines the accuracy with which E_H can be calculated. Substituting a_0 gives the value of the Rydberg constant $\overline{R}_\infty$ as given in Eq. (2)

$$\overline{R}_\infty = \frac{E_h}{2hc} = 109,737.31568527(73)\,\text{cm}^{-1}.$$

To convert E_H from J to wave numbers, we divided by hc, where c is the speed of light, which is defined to be 299,792,458 m s^{-1}.

However, the nucleus is not an infinite mass, and it is common practice to acknowledge the fact that the center of mass of the atom does not quite coincide with that of the nucleus. To take this into account, the reduced mass, μ, of the system proton p and electron e is defined as

$$\frac{1}{\mu} = \frac{1}{m_p} + \frac{1}{m_e}$$

This means that the $\overline{R}$-value of the dynamic system is found from

$$\frac{R_\mu}{\overline{R}_\infty} = \frac{\mu}{m_e} = 0.9994556794$$

using the IUPAC (2007) values [2] for m_p and m_e. The theoretically calculated value of $\overline{R}$ with respect to the reduced mass μ is given by

$$\overline{R}_\mu = \overline{R}_\infty\left(\frac{\mu}{m_e}\right) = 109,677.583\,\text{cm}^{-1}.$$

The values of the constant of proportionality, $\overline{R}$, found from *atomic spectral series* must be compared with this theoretically calculated value of $\overline{R}_\mu$. It is shown below that this value is about 1.25 cm^{-1} *smaller* than that found from the Lyman UV series of atomic spectral lines, namely, 109,678.7717 cm^{-1}. Attention is also drawn to the fact that the theoretically calculated value of $\overline{R}_\mu$ does not agree with the *experimental ionization energy* 109,678.7717 cm^{-1}. However, the experimental ionization energy agrees within experimental error with the experimental values found for $\overline{R}_{\text{intercept}}$, as discussed below. This is further taken up in Sect. 2.3.

2 Some Graphing Background

2.1 Straight Lines

In this section, a brief account is given of the usual plotting conventions that atomic spectroscopists use to plot and interpret their experimental data, since it is necessary for the understanding of the discussions below, especially that of Sect. 3.

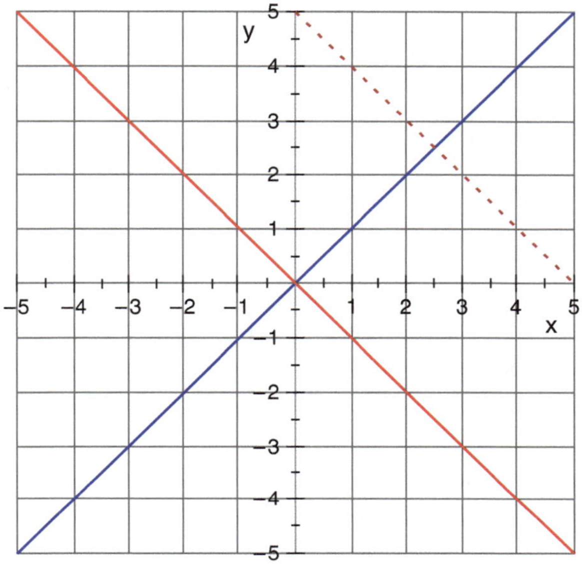

Fig. 1 Plot of the two straight lines $y = \pm x$. For the *red lines*, the *negative sign* holds, and for the *blue line*, the *positive signs*. The part of the straight line in the fourth quadrant is shifted upward by $y = E/\overline{R} = 5$ units as an example; this convention is normally used in atomic spectroscopy

If we define new variables $y_n = E_n$ and $x_n = \frac{1}{n^2}$ and substitute them into Eq. (1), then the fundamental (hyperbolic) relation reduces to that of a straight line

$$y_n \sim -x_n$$

The equations $y = \mp x$ are drawn in Fig. 1, assuming that the variables are continuous variables and that the constant of proportionality is equal to one, using the color codes described in the legend. For purposes of atomic spectroscopy, only the fourth quadrant part of the red curve is used. It is the usual practice of atomic spectroscopists to shift this curve upward, so that it is placed in the 1st quadrant, as shown by the dotted line in Fig. 1. The actual upward shift depends upon the specific case, but in the case of atomic spectra, it is the usual practice to shift the straight line upward by adding $\overline{R}$ to the energy values themselves, that is, $-E_\infty = 0$ now becomes $E'_\infty = \overline{R}$, and all the other energy points are now positive and lie in quadrant 1; this procedure is used for Figs. 2 and 3.

2.2 Lines Through Discrete Points

The linear relationship of Eq. (1) implies that $\lim_{n\to\infty} E_n = 0$. This means that the linear graph passes through the origin. Since the energy of a bound state is defined

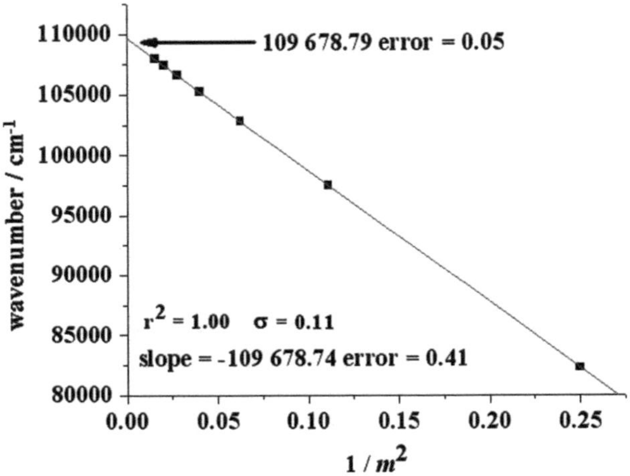

Fig. 2 The plot of Eq. (5) for the Lyman vacuum-UV series of the hydrogen atom for which the quantum number of the final state is $n = 1$. See discussion in the text. The data are from Sansonetti [8]

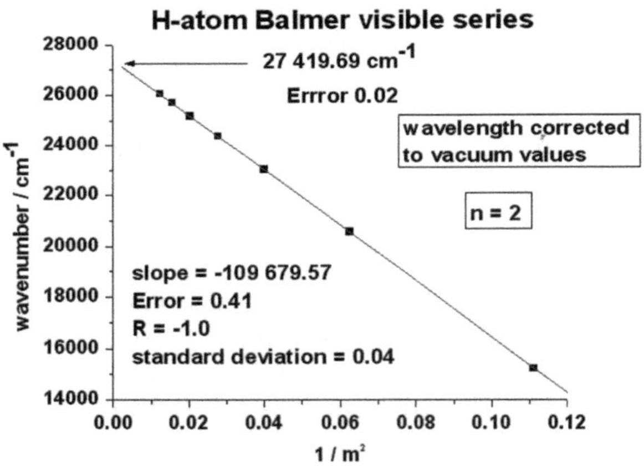

Fig. 3 The plot of Eq. (4) for the Balmer visible series of the hydrogen atom for which the quantum number of the final state is $n = 2$. The wavelength values were converted from air to vacuum following Lide [9]. See discussion in the text. The data are from Sansonetti [8]

to be negative, the linear graph of E_n vs $\frac{1}{n^2}$ lies in the *fourth quadrant* as explained above, having a negative slope $-\overline{R}$. It is, however, usual in atomic spectroscopy to add the energy $\overline{R}$ to each energy value E_n, so that the graph is, as it were, shifted upward by $\overline{R}$ to appear in the *first quadrant*. The intercept on the energy axis is then at $\overline{R}$, but the slope has not changed; this methodology is used in Fig. 2.

It has become customary to say that any energy value E_m to the *left* of any other energy value E_n on the graph is called a *higher energy value* or even a *higher energy level*; this implies that $m > n$.

The concept of "linear graph" used above is actually a misnomer, because Eq. (1) does not refer to the graph of a continuous function but to a discrete integer relationship: There are no energy values between any two adjacent discrete points n_i and $(n_i \pm 1)$ on the graph. The graph only consists of a set of *discrete points*, although for the better visualization of the implied linear relationship described by Eq. (1), a straight line is usually put through them, especially for statistical purposes.

Whichever plotting method is used, it is clear from Eq. (1) that the energy difference between points E_m and E_n for *integral* $m > n$ is given by the expression

$$\Delta(E_m - E_n) = \Delta E_{m\to n} = \Delta E_{mn} = -R\left\{\frac{1}{m^2} - \frac{1}{n^2}\right\} = R\left\{\frac{1}{n^2} - \frac{1}{m^2}\right\}. \qquad (3)$$

Since the integral number m refers to a *discrete higher energy state* and n to a *discrete lower state*, this expression represents a *transition* of the H-atom from a state described by the quantum number m to a state described by the quantum number n, emitting a quantum of energy

$$\Delta E_{mn} = h\nu_{mn}\mathrm{J}$$

where h (J s) is Planck's constant and ν (s^{-1}) the frequency of the monochromatic emitted light; ΔE_{mn} and R are measured in J.

It is usual in atomic spectroscopy to rewrite Eq. (3) in terms of the energy unit of *wave number*, $\overline{\nu}$ (cm^{-1}), by dividing both sides of Eq. (3) by hc, where c is the speed of light in units of cm s^{-1}, thus obtaining

$$\overline{\nu}_{mn} = \overline{R}\left\{\frac{1}{n^2} - \frac{1}{m^2}\right\} \qquad (4)$$

where $\overline{R}$ *is given in wave numbers*; unfortunately, in many cases, the same symbol R is used for the values of the proportionality constant given in either J or wave numbers, but usually no confusion can arise. The *Rydberg–Ritz equation* as written in Eq. (4) is used in atomic spectroscopy.

2.3 *Specializing to H-Atom Spectroscopy*

Accepting the validity of Eq. (1) and thus of Eqs. (3) and (4)—whether derived theoretically or empirically from atomic spectroscopy—any electronic transition in the hydrogen atom between a *discrete higher* energy state, E_m, and a *discrete lower* energy state, E_n, is described in terms of the discrete energy difference, ΔE_{mn}, shown in Eqs. (3) and (4).

Some reflection shows that a least-squares straight line can be fitted through the discrete points of Eq. (4) for $n = constant$, that is, all transitions in an atomic line series toward the *same final state* E_n will have discrete energies (wave numbers) according to the linear relationship

$$y = b - ax \tag{5}$$

where

$$y = \overline{\nu}_{mn}; \quad x = \frac{1}{m^2}; \quad a = \overline{R}; \quad b = \overline{R}\frac{1}{n^2} \tag{6}$$

for $m > n$ where $n = constant$.

Equations (5) and (6) state that the *slope* a of least-squares line drawn through the discrete points of a hydrogen-atom line series and its *intercept* b on the energy axis are related by the multiplicative factor $\frac{1}{n^2}$, that is, it follows that

$$a = n^2 b \tag{7}$$

where n is the quantum number of the final (lower) state at which the transitions of each of the lines in the series terminate.

Attention is drawn to the fact that if a good straight-line fit for the *experimental values* of $\overline{\nu}$ is obtained, such that $r^2 = 1$, where r is the variance, then it is usually assumed that any theoretical model that reproduces Eqs. (1) and (3) is supported by "experimental evidence." In such a case, it is usually assumed that no adjustments need to be made by massaging either the theory or the experimentally measured wave numbers. However, it is pointed out that such a fitting procedure only says that there *seems to be* a relationship between the energy of a transition and the difference between the inverse-squared quantum numbers as expressed in Eqs. (5) and (6). *Even a good fitting does not imply that*:

- Both values of $\overline{R}$ obtained from the fit should be identical within experimental error to the theoretically calculated value of $\overline{R}_\mu$.
- The value of $\overline{R}$ obtained from the *slope* of the line and that obtained from the *intercept* on the energy axis should necessarily be identical.

Both of these very important issues are discussed below.

Equation (7), which implies the equivalence of the $\overline{R}$-values as calculated from either the *slope* of the least-squares line of any series of atomic lines of the H-atom or from its *intercept* on the energy axis, is very important. In fact, it sets a stringent condition for the validity of Eq. (1) and hence, also for Eq. (3):

> If and only if the value of $\overline{R}$ found from the slope and the value found from the intercept on the energy axis are identical within experimental uncertainties, then it can be concluded that the inverse-squared relationship of Equation (1) is satisfied by the experimental data.

If the condition set by Eq. (7) for the experimentally determined values of $\overline{\nu}$ in any hydrogen-atom line series is *not fulfilled* for any one or for all the measured hydrogen-atom spectral line series within the accuracy of the linear least-squares fit, then there is something fundamentally wrong with Eq. (1).

For the same reasons, it can then also be concluded that any theory that assumes the correctness of the inverse-squared relationship exemplified by Equation (1) is inadequate.

It furthermore follows that the postulates which led to the theoretical derivation of Equation (1) must be fundamentally revised.

Attention is especially drawn to the fact that any procedure or procedures that "modify" or "massage" the experimental data so that they conform to the stringent conditions of Equation (7) and hence to Equation (1) *do not cure the underlying inadequacies in the fundamental model postulates that led to Equation* (1). Such procedures only serve to obscure the inadequacies of the theoretical model and are used to enhance the *belief* that the "theory is correct and experimentally proven."

2.4 The Various Proportionality Constants $\overline{R}$

Attention is drawn to the fact that *four different proportionality constants* $\overline{R}$ are used in this chapter as well as the experimental ionization energy. These constants, which are all measured in wave numbers in this chapter, are:

1. $\overline{R}_{\infty}$, the proportionality constant *calculated theoretically* with respect to infinite mass of the nucleus from any of the (nonrelativistic) models of the hydrogen atom.
2. $\overline{R}_{\mu}$, the proportionality constant *calculated theoretically* with respect to the reduced mass μ of the hydrogen proton and electron.
3. $\overline{R}_{\text{slope}}$, the proportionality constant obtained from the *slope* of a linear plot of the *experimental data* for a specific line series of the hydrogen atom.
4. $\overline{R}_{\text{intercept}}$, the value of the proportionality constant obtained from the *intercept* on the energy axis of a linear plot of the *experimental data* for any hydrogenic atom atomic spectral line series.
5. I, the *experimentally measured ionization energy* of the single electron of a hydrogenic atom or ion.

It is emphasized that Eq. (1) can only be assumed to be correct when $\overline{R}_{\text{intercept}} = \overline{R}_{\text{slope}} = I$ within experimental error. It is not adequate when only $\overline{R}_{\text{intercept}} = I$.

The first two points are discussed in Sect. 1, while the third and fourth are discussed in Sect. 2.3; they are determined from experimental data in Sect. 3. The values of the fifth are used in Sect. 3.

From Sect. 1, it is clear that for the *theoretically calculated entities* $\overline{R}_{\infty}$ should be larger than $\overline{R}_{\mu}$. It is, furthermore, the ideal that the value of $\overline{R}_{\mu}$ should be as close as possible to the experimentally determined value of $\overline{R}_{\text{slope}}$ and $\overline{R}_{\text{intercept}}$ for any theory to be able to claim that it is verified by experimental results. In this chapter, we show in Sect. 3 that it is not the case.

The theory developed in Sect. 2.3 and applied in Sect. 3 shows that if the fundamental relation of Eq. (1) is correct, then the experimental plots of the line series of the hydrogen atom should give proportionality constants such that $\overline{R}_{\text{slope}} = \overline{R}_{\text{intercept}}$ within experimental error. It is shown in Sect. 3 that this is not the case.

In addition, it is shown in Sect. 3 that the errors in the linear least-squares fit of the straight line through the experimental frequencies of any hydrogen-line spectra should be the same for $\overline{R}_{\text{slope}}$ and $\overline{R}_{\text{intercept}}$, since they stem from the same data set if the fundamental relation of Eq. (1) is correct. In Sect. 3, it is shown that this is not the case.

Finally, the experimental ionization energy I measures the energy required for an electron to ionize off from any hydrogenic atom nucleus, such as ${}^{6}_{3}\text{Li}^{2+}$. This means that the following relationship should hold for the experimental data, namely, $\overline{R}_{\text{slope}} = \overline{R}_{\text{intercept}} = I$, as discussed in Sect. 3. If the assumption of the reduced mass in the quantum-mechanical model of a hydrogenic atom is adequate, then the model should thus yield an expression for $\overline{R}_{\mu}$ such that it equals the experimental ionization energy of that atom. In Sect. 3, it is shown below that this is not the case. This means that the value of $\overline{R}_{\infty}$ does not agree with the experimental ionization energy.

A huge effort is made by theoreticians to explain the discrepancies between the experimental values and the theoretical values, based upon the fundamental relation of atomic spectroscopy. This rather complicated process to calculate "corrections" that can be applied to experimental results to duplicate the experimental value of $\overline{R}_{\infty}$ is briefly summarized in Sect. 4. However, it is shown below that the origin of the problem is found in the fact that the fundamental relation of atomic spectroscopy is almost but not quite adequate to agree with the experimental facts. No amount of corrections can salvage an incorrect equation.

3 Atomic Spectroscopy of the H-Atom

3.1 Introductory Remarks

As said above, Eq. (1) is used as an *empirical working postulate* to "explain" the occurrence of the line series of atomic spectra and their frequencies (or wavelengths or wave numbers, which are the same thing) as found *experimentally*; this is the approach first used by Balmer in 1885 to explain the Balmer series of the hydrogen atom. This *empirical working postulate*—as shown in Sect. 2—can be used as an *internal check* on its own validity as well as a check on the validity of any nonrelativistic theory which produces Eq. (1) from the application of its model postulates. In addition, the empirical equation can be used to check whether the values of $\overline{R}$ as derived empirically and $\overline{R}_{\infty}$ as calculated theoretically and given by Eq. (2) are identical within experimental error.

Table 1 The high-resolution data for the Lyman series of the hydrogen atom, for which $n = 1$

m	$1/(m^2)$	λ (Å)	$\overline{\nu}$ (cm^{-1})
2	0.25000000	1,215.67364	82,258.919425
2	0.25000000	1,215.66824	82,259.284819
3	0.11111111	1,025.72220	97,492.283973
4	0.06250000	972.53670	102,823.883150
5	0.04000000	949.74300	105,291.642055
6	0.02777778	937.80340	106,632.157657
7	0.02040816	930.74820	107,440.444150
8	0.01562500	926.22560	107,965.057325

The experimental data are from Sansonetti [8], which are fitted to Eq. (5), remembering that $\overline{\nu} = \Delta(E/hc)$

3.2 *Testing the Lyman Series of H*

For the Lyman ultraviolet series of the hydrogen atom, with $n = 1$, Eq. (5) reduces to

$$y = \overline{R} - \overline{R}x \tag{8}$$

since $b = a$ in Eq. (7). A plot of E_m of the lines in the Lyman series (in wave numbers) is plotted against the discrete values of $x = \frac{1}{m^2}$ for $m > 1$ should display a linear relationship. A least-squares straight line drawn through the discrete points should show that the *slope* of the line and the *intercept* on the energy axis must be identical to within experimental uncertainties. In fact, they both must be equal to $\overline{R}$.

> This is a rigorous requirement, and it serves as an absolute test for the validity of Eq. (1), independent of whether it is established empirically, or follows as a result of a theoretical model.

This statement still says nothing about any relationship between the experimentally derived value of $\overline{R}$ and the calculated value of $\overline{R}_\infty$ as accepted by CODATA and IUPAC (which is described in [1–8]); this is again taken up in Sect. 4.

Very accurate high-resolution vacuum-UV data with an error of 0.000 1 Å for the Lyman series of the line spectrum of the hydrogen atom are reported in [8]. This error, translated into wave numbers, constitutes an error of about 0.01 cm^{-1} in the wave numbers around 100,000 cm^{-1}.

The atomic line-spectral data for the Lyman vacuum-UV series of the hydrogen atom are given in Table 1. The plot of the line wave number vs $1/m^2$, following Eqs. (7) and (8), is displayed in Fig. 2. The least-squares fitting is excellent, with $r^2 = 1.000$ and $\sigma = 0.41\ cm^{-1}$. The values obtained for the constant of proportionality from the least-squares fitting are:

$$\overline{R}_{\text{intercept}} = 109,678.79 \pm 0.05\ \text{cm}^{-1}$$

$$\overline{R}_{\text{slope}} = 109,679.57 \pm 0.41\ \text{cm}^{-1}$$

$$(\overline{R}_{\text{slope}} - R_{\text{intercept}}) = 0.78\ \text{cm}^{-1}.$$

Formally this least-squares fitting shows that $b \neq a$. However, the difference between the two values of $\overline{R}$ is just smaller than twice the largest error obtained in the least-squares fitting of the experimental data. This is a very small difference, and superficially the conclusion may be reached that this fitting agrees with that expected from Eqs. (7) and (8). Howsoever small this difference is, it lies at the upper end of the range for which it can with confidence be stated that the linear least-squares fitting supports the requirement that the two values of $\overline{R}$ should agree within experimental error. Another problematic aspect of the linear least-squares fitting is that the *error* in the value of $\overline{R}$ calculated from the intercept is eight times smaller than that calculated from the slope, although both *follow from the same set of experimental data and the same fitting process*; this is discussed in more detail below in Sect. 3.4.

The values obtained for the constant of proportionality $\overline{R}$ are, respectively, 58.53 cm^{-1} (intercept) and 57.75 cm^{-1} (slope) *smaller* than the value of the Rydberg constant, $\overline{R}_{\infty}$, as given in Eq. (2), although both of them are close to the value of $\overline{R}_{\mu}$. The numerical differences between the values of the constants calculated from the slope and the intercept are larger than that which can be expected from any experimental or least-squares fitting error, and it, together with the uncertain conclusion about the mutual agreement of the two values of $\overline{R}$ discussed above, places question marks over the validity of Eq. (1), which is the fundamental equation of empirical atomic spectroscopy. The same applies to the IUPAC and CODATA value of the Rydberg constant. It is, however, the usual practice in atomic spectroscopy to assume corrections that have to be made to theory to bring the values of the constant of proportionality, $\overline{R}$, and that of the Rydberg constant, R_{∞}, into coincidence. Such corrections are, for instance, the mutual effect of the motions of the electron and proton on one another, in which both are considered to be *particles*, the interaction of the emitted photon with the vacuum, etc. It is our considered opinion that the *raw data* must suffice, if not for the theoretical value of the Rydberg constant, then certainly for the empirical Eq. (1).

3.3 Testing the Balmer Series of H

The Balmer series, for which the final state is characterized by $n = 2$, occurs in the visible range of the electromagnetic spectrum. The wavelengths of its line spectrum are usually measured in air, and a correction is needed to convert them to vacuum wavelengths. This is done according to the interpolation table given by Lide [9], and the resulting data are plotted in Fig. 3; the straight line shows the least-squares fitting through the discrete points. Again, the correlation coefficient and the standard deviation show that the least-squares fitting indeed visually indicates a linear relationship.

In comparing the values of the constants of proportionality, $\overline{R}$, we refer to Eq. (7), that is, four times the intercept on the energy axis must give the value of $\overline{R}$, while the slope gives the value of $\overline{R}$ without any multiplicative factor. These are found as

$$\overline{R}_{\text{intercept}} = 109,678.76 \pm 0.02\,\text{cm}^{-1}$$
$$\overline{R}_{\text{slope}} = 109,679.57 \pm 0.41\,\text{cm}^{-1}$$
$$(\overline{R}_{\text{slope}} - R_{\text{intercept}}) = 0.81\,\text{cm}^{-1}.$$

Again, their difference is just less than that of twice the largest error, again leading to an uncertain decision about whether they are identical or not. However, if we consider the values of the respective constants of proportionality derived above for the Lyman series and those of the Balmer series derived here, it is clear that they are identical, and their respective differences are also just about equal too. It is highly unlikely that any systematic trends in the respective experimental data for the two sets, combined with uncertainties about the linear least-squares fitting, would give two almost identical sets of values for $\overline{R}_{\text{slope}}$ and $\overline{R}_{\text{intercept}}$. In this case, the error in the intercept is about *20 times smaller* than the error in the slope; this phenomenon is discussed in Sect. 3.4.

3.4 Errors in the Slopes

There is one more very important factor to consider, namely, the *large errors* (~8 to ~20 times larger, respectively) found for the *slopes* of the Lyman and Balmer line series of the hydrogen atom compared to the *small errors* found for the *intercepts* on the energy axis. The slope of the least-squares straight line is computed from an energy difference divided by the difference in the respective values of $1/m^2$, that is, from

$$\left(\frac{\Delta E_m}{\Delta x}\right) = \frac{(E_{m_1} - E_{m_2})}{\left(\frac{1}{m_1^2} - \frac{1}{m_2^2}\right)} \tag{9}$$

where m_1 and m_2 are two discrete points belonging to the energies E_{m_1} and E_{m_2}, respectively. Theoretically, the denominator in this equation should not contribute to any errors, since it is derived from *defined discrete quantities*. This, in turn, means that the errors in the respective slopes are *solely due to experimental errors in the energies* (wavelengths), or in other words, to their differences, $(E_{m_1} - E_{m_2})$. However, we are working with very accurate wavelength data for the Lyman series (accurate to within 0.000 1 Å, thus introducing an error of not larger than 0.01 cm^{-1} at around 1,000 Å), as well as with accurate data for the Balmer series (up to at least 0.001 Å). The source of the errors in the slope are thus almost certainly not only due to inaccurate data, and it must lie in the inverse-squared assumption of Eq. (1).

Any discrepancies between the experimental data and the assumption of Eq. (1) will thus be magnified for the slope and, especially, for its least-squares errors. Since our data are far more accurate than the errors in the slopes imply, the conclusion follows that the *large errors* in the values of the slopes must originate in the fact that the differences, Δx, in the respective values of $(1/m^2)$ are not as accurate as defined.

In fact, it should actually be somewhat *smaller* than those predicted by Eq. (5), thus causing a decrease in the least-squares errors for the slope. If this argument were true, it means that the values of $\overline{R}$ found from the slopes and intercepts are definitely different, thus leading to the conclusion that the inverse-square relation of Eq. (1) as expressed in Eq. (4) is not exact.

The problem facing us now is as follows: Just where does the discrepancy with the defined values of Eq. (1) originate? Should we massage the experimental energy values, or should we search for a modified inverse-squared relationship? Or both? Atomic spectroscopy usually chooses the first option, intensely massaging the data by empirically added corrections to the energy values till the correct value of $\overline{R}_\infty$ is obtained. We prefer the second option: there is something essentially inadequate in the models underpinning the present quantum-mechanical system of thought.

3.5 Testing the Lyman Series of Li-III

Finally, let us look at the Lyman vacuum-UV line spectrum of the *hydrogen-like* Li-III entity, that is, the doubly ionized lithium atom. Its line spectrum is also described by transitions to the ground state with $n = 1$, but the nuclear charge is different from that of the hydrogen atom. Therefore, we add a multiplicative factor $Z^2 = 9$ to the equations. In effect this means that the values of the intercept on the energy axis and that of the slope as determined by the least-squares regression analysis must be divided by 9 to obtain the respective values of $\overline{R}$. Such a plot, using the data from Moore [10], where we call the constant of proportionally K, gives the following for the energy cutoff and the slope:

$$\begin{aligned} K_{\text{intercept}} &= 987,497.39 \pm 159.28\,\text{cm}^{-1} \\ K_{\text{slope}} &= 986,753.61 \pm 1,473.57\,\text{cm}^{-1} \\ (R_{\text{slope}} - R_{\text{intercept}}) &= 743.78\,\text{cm}^{-1} \end{aligned}$$

whence, by dividing by 9, we find

$$\begin{aligned} R_{\text{intercept}} &= 109,721.93 \pm 17.70\ \text{cm}^{-1} \\ R_{\text{slope}} &= 109,639.29 \pm 163.73\ \text{cm}^{-1} \\ (R_{\text{slope}} - R_{\text{intercept}}) &= 82.64\ \text{cm}^{-1}. \end{aligned}$$

The difference between the $\overline{R}$-values found from the slope and the cutoff is much larger than with hydrogen, although it is much less than the largest error, namely, that of $\overline{R}_{\text{slope}}$. It is interesting to compare the calculated values of $\overline{R}_\mu$ for the two lithium isotopes with respect to the reduced masses $\mu({}^6_3\text{Li}) = 5.48429818 \times$

10^{-4} u and $\mu({}^{7}_{3}\text{Li}) = 5.484620264 \times 10^{-4}$ u. These are, respectively, $\overline{R}_{\mu}({}^{6}_{3}\text{Li}) = 109,707.2914\,\text{cm}^{-1}$ and $\overline{R}_{\mu}({}^{7}_{3}\text{Li}) = 109,713.7351\,\text{cm}^{-1}$, which agree fairly well with that of the intercept found above, thus underlining the trend found for hydrogen.

It is usually argued that this occurs because we are using the "bare" nuclear charge here, and that in practice, this charge is "partially screened by the single electron", although the physical mechanism of a single electron, as it were, screening itself, is not clear. In order to compensate for the large discrepancy, a nuclear-charge screening factor p is *empirically* brought in to reduce the nuclear charge to an *effective nuclear charge* $(Z - p)$. This implies that the constant K of the above expressions must be divided by $(Z - p)^2$, thus making $\overline{R}_{\text{intercept}}$ larger; by manipulating p, it is possible to make $\overline{R}_{\text{intercept}}$ large enough to coincide with that expected from $\overline{R}_{\infty}$. This looks as though it might "work" and raises the value of $\overline{R}_{\text{intercept}}$, but apart from the fact that there is no real explanation of why this particular value of p should be introduced, it would make the uncertainties in both values of $\overline{R}$ even larger; in addition, the same value of p will *never* bring $\overline{R}_{\text{slope}}$ and R_{cutoff} to coincide.

To conclude, the large error found for $\overline{R}_{\text{slope}}$ can be explained in the same way as for the hydrogen cases discussed above: *the inverse-squared law is almost but not quite exact.*

3.6 Spectroscopic Conclusions

Equation (1) is contained in the relationship represented by Eq. (4), which was *empirically* postulated for $n = 2$ by Balmer in 1885, based upon the spectral data known at the time.

In our description of the phenomenon of the line series of hydrogenic atoms, based upon accepting the validity of Eqs. (1) and (4), we showed that every linear regression plot of the experimental data of a single line series of the H-atom should yield *two values of the so-called constant of proportionality*, the one coming from the *slope*, $\overline{R}_{\text{slope}}$, and the other from the value of the *intercept*, $\overline{R}_{\text{intercept}}$, on the energy axis. These two constants are related by Eq. (7) for the ideal case where Eqs. (1) and (4) exactly fit the data. In the case of the *Lyman series*, these two values should be identically equal; for the *Balmer series*, the value from the intercept should be multiplied by 4 to obtain the same value as that obtained from the slope. For the *Lyman line series* of Li-III, the nuclear charge $Z^2 = 9$ is added, so that $9a = 9b$, whence it follows that the two values of $\overline{R}$ should be identical.

Our data today are much more accurate than those available to Balmer, and when we use them, it is clear that the relationship of Eq. (4) does not quite apply to the line series of the one-electron atomic entities. Our analysis pointed out four facts from the linear regressions:

1. The constants of proportionality, which we derive from a linear least-squares regression fit of the data, *are not equal*, as required by Eq. (7), that is, $\overline{R}_{\text{intercept}} \neq \overline{R}_{\text{slope}}$. In fact, the difference between them is larger than warranted by experimental uncertainties.
2. The errors in the $\overline{R}$-values derived from the *slopes* are far larger than can be expected from experimental errors. In addition, they are far larger than those of the *intercepts* on the energy axis.
3. The introduction of the nuclear charge does not remove the problem of the constants of proportionality, since the $\overline{R}$-values of the Li-III entity determined from its Lyman wavelengths are also not identical.
4. In all cases examined, the values of $\overline{R}$ for the *intercepts* for the respective cases almost but not quite agree with calculated values of $\overline{R}_{\mu}$, as well as with the respective experimental values of the ionization energies.

These four points taken together show clearly that the inverse-squared law of Eqs. (1) and (4) is *inadequate* to describe the experimental line spectra of one-electron atoms, since they "fail" the test formulated in Sect. 2, as shown in Sect. 3. This means that either the energies of the states (i.e., the wavelengths of the lines) or the inverse-squared parts are not quite correct, or both.

The *experimental ionization energy* for the hydrogen atom can be found from the intercept of the linear regression line for the Lyman and Balmer series. For both series, the intercept agrees with the experimental ionization energy in its ground electronic state, namely, 13.598433 eV = 91.17534638 nm = 109,678.7717 cm^{-1}, compared with the Lyman value of $\overline{R}_{\text{intercept}} = 109{,}678.79 \pm 0.05$ cm^{-1} and the Balmer value of $\overline{R}_{\text{intercept}} = 109{,}678.76 \pm 0.02$ cm^{-1}. This gives perfect agreement between the experimental ionization energy and the regression values of $\overline{R}_{\text{intercept}}$ found for hydrogen.

The experimental ionization energy of the Li-III entity is 122.45429 eV = 987, 660.5864 cm^{-1}, while the value from the intercept on the energy axis is $K_{\text{intercept}} = 987{,}497.39 \pm 159.28$ cm^{-1}, which can be considered to be equal to the K-value, although it just falls outside the error boundaries indicated. The experimental ionization energies used here are from Lide [9, pp.10–205]. It must be pointed out that Moore [10] gives two values of the ionization energy, namely, 122.419 and 122.420 eV = 987,384.02 cm^{-1}; both these values are within the error bounds for the linear regression.

> The weight of the evidence presented in this analysis of the hydrogen and Li-atoms' atomic spectral data indicates that the inverse-squared relationship of Equation (1) is very nearly but not quite correct.

All theoretical treatments leading to the inverse-squared relationship expressed by Eq. (1) suffer from the malady that they cannot predict the fine structure found in the high-resolution atomic spectra of the hydrogen atom. This means that many different, in some cases, empirical corrections have to be applied to the inverse-squared values of the hydrogen lines [6]. All these calculations and additive corrections are based upon a space with coordinates in which time can be split off from the three Cartesian space coordinates, which could be the cause of the deviations from the inverse-squared ideal solution found here.

4 Empirical Corrections for the H-Atom

4.1 Attempts to Correct the Quantum Numbers

Referring to Eqs. (1)–(4), we point out that the main quantum number is *integral*, leading to *discrete* energy states. We stress the use of the words *integral* and *discrete* here because $\overline{R}$ is a proportionality constant and the *main quantum numbers are integral numbers*, which cannot be empirically modified. Any attempt to modify Eq. (4) by fitting it to empirical equations in which the quantum numbers are considered to be *nonintegral variables*, such as

$$\overline{\nu}_{m\to n} = \overline{R}\left(\frac{1}{(n-\delta_b)^2} - \frac{1}{(m-\delta_a)^2}\right) \tag{10}$$

to obtain corrective constants δ_b and δ_a in order to force the equivalence of the constants $\overline{R} \to \overline{R}_\infty$, destroys the fundamental assumption of the *quantum nature of the process of emission*. It actually introduces the very strange concept of *fractional or effective quantum numbers*, such as $(n - \delta_b)$, which cannot ever be fitted into the scheme by which the *discrete main quantum numbers* were established by Schrödinger and Heisenberg for nonrelativistic quantum mechanics. This type of equation is written in a way that leaves the "comforting feeling" that it is *still* a quantum mechanically derived equation involving two true quantum numbers, but it is actually completely empirical. In fact, it is the same type of equation that appears in the section dealing with electrodynamics below, where the value of the constant $\overline{R}$ for each transition with wave number $\overline{\nu}_{m\to n}$ is found from the empirical expression

$$\overline{R}_{m\to n} = \overline{R}_\infty \times (f_{m\to n}) \tag{11}$$

where the factor $f_{m\to n}$ is obtained by applying successive layers of complicated quantum-electrodynamical *calculations*; naturally, other quantum numbers also play a role in this process.

4.2 Quantum-Electrodynamical Contributions

Some of the discrepancies between the experimental values of the line spectrum of the H-atom have been noticed before, and elaborate corrective schemes have been devised to bring experiment and theory into agreement [1–7]. We quote from [7]:

Quantum electrodynamics (QED) makes extremely accurate predictions despite the 'mathematical inconsistencies and renormalized infinities swept under the rug' [11]. With the assumption that the theory is correct, it is used to determine values of the relevant physical constants by adjusting their values to give the best agreement to experiments [5]. The CODATA recommended value of the Rydberg constant has been obtained primarily by comparing theory and experiment for 23 transition frequencies or pairs of frequencies in hydrogen and deuterium. The theoretical value for each transition is the product of the Rydberg constant and a calculated factor based on QED that also depends upon other constants. $\cdots$ the recommended value of the Rydberg constant has a larger relative uncertainty of 6.6×10^{-14} which is essentially the uncertainty of the theoretical factor. The main source is the uncertainty in the charge radius of the proton with additional uncertainty due to uncalculated or partially calculated higher-order terms in the QED corrections.

In all fairness to this approach, it is pointed out that Feynman, one of the main developers of this field of quantum science, shared the misgivings of Dyson [11] expressed above. From very accurate modern experimental data, it is possible, after applying adequate and self-consistent corrections due to the *Lamb shift*, to obtain the value of the Rydberg constant at least as accurate as that of its theoretically calculated value according to Eides, Grotch, and Shelyuto [4]. The accuracy of the experimentally determined Rydberg constant depends upon three variables, namely, (a) the accuracy of the frequency determination, (b) the accuracy of the self-consistent factor due to the Lamb shift, and (c) the accuracy of the electron–proton mass ratio. For some methods, (d) the accuracy of the proton radius also plays an important part. The largest errors are introduced by factors (a) and (b). The best value of the experimental Rydberg constant as listed by [4] is $109{,}737.315686(9)\,\text{cm}^{-1}$, which is comparable to the IUPAC value of $109{,}737.31568527(73)\,\text{cm}^{-1}$. The *Lamb shift* is a small splitting between the $^2S_{\frac{1}{2}}$ and the $^2P_{\frac{1}{2}}$ levels of the hydrogen atom—in contradistinction with the prediction of the Dirac theory of the hydrogen atom, which does not take the interaction with nuclear moment into account. This shift is explained in terms of quantum electrodynamics. It is assumed that there is a fluctuation in the electric and magnetic fields, which are associated with the concept vacuum, as well as the exchange of virtual photons. This, in turn, perturbs the Coulomb potential governing the motion of the electron around the nucleus. This perturbation affects the position of the electron, thus causing the splitting in the levels. The splitting of these levels are experimentally found to be about $1{,}040\,\text{MHz}{\sim}0.03469\,\text{cm}^{-1}$.

4.3 Flaws of Empirical Corrections

Per-Olov Löwdin of Uppsala used to start his summer school lectures in quantum theory with the words:

Experiment comes first, second and third. Then comes theory!

It is, therefore, essential to measure the predictions of any theory against experimental data. If discrepancies are found between the numerical predictions of quantum mechanics as postulated by Schrödinger and Heisenberg and the best available experimental data, should one rather, as it were, go back and *improve the original model assumptions*, that is, the theory must be reinvented, or should one rather search for one or more *external "corrective assumptions"* that can be applied in a *post hoc* fashion to the theory to "improve" the agreement between theory and experiment? Up to now, the second option was consistently chosen by quantum chemists, and today there are a plethora of such corrective methods available, some of them even adding corrections to corrections to corrections in a cascading fashion (e.g., those called first-order corrections, second-order corrections). Moreover, such corrections are usually motivated in a language that "sounds quantum mechanical" and thus belies their essential empirical nature—and mostly also their purely classical nature. The parade-horse example of this phenomenon is the electron spin and its associated half-integral quantum number, which is empirically *grafted on* to the quantum mechanics of Heisenberg and Schrödinger (it does appear as a fundamental result in the Dirac theory of the hydrogen atom, although it is pointed out that this is rather inevitable, since Dirac *a priori assumed* the Pauli spin matrices), and even today, the electron spin forms an indispensable part of every quantum-mechanical computational program in the form of antisymmetric wave functions appearing in symmetrized linear combinations of atomic orbitals (LCAO).

The very important question now is: Can the result of such a *post hoc* empirical and/or classical corrective methodology *still* be said to be quantum mechanical or even "based upon" quantum mechanics? It is our considered opinion (see Boeyens and Schutte [12]) that the answer to this question is a very clear "No!" Basically, the same response was formulated by Schopenhauer in his aphorism (freely translated):

> When one has a pail of pure milk and a pail of sewage, and one adds one drop of pure milk to the pail of sewage, then one has sewage, and when one adds one drop of sewage to the pail of pure milk, one also has sewage.[2]

The analogy is clear when *sewage* is rather irreverently in this context identified as empiricism and/or classical physics and *pure milk* as "pure" quantum mechanics according to the original assumptions: The *post hoc* addition of corrections to theoretical quantum-mechanical results, whether purely empirical and/or classical, *invariably* turns the result into something that does not anymore rest on the original set of assumptions.

I have examined the corrections listed above, and I think that all of them fail the Schopenhauer test: they adulterate the quantum-mechanical methodology as defined above. There is thus a need to go back to the fundamental assumptions of quantum mechanics and *reinvent* them to produce results that coincide with experimental data. I realize that this is going to be a difficult process, since the methods we criticize here have been part of the teaching of all aspects of the quantum theory for about 80 years.

[2]The "sewage" in the original German was called "*Dreck*."

5 On the Signs of E and $\frac{1}{n^2}$

It is of interest to further discuss the signs of the variables in Eq. (1), which are given as energy units for E, the same energy units for R, while n is a pure number, the quantum number. Introducing a new variable $K_n = \frac{E_n}{R}$, we can write the equation in dimensionless units as

$$K_n = -\frac{1}{n^2} \tag{12}$$

As described above, it is the usual practice to plot the energy E_n against $\frac{1}{n^2}$ to obtain a straight line in the fourth quadrant with slope -1 as explained in Sect. 2. This line passes through the origin when $\lim n \to \infty$, indicating that the system has reached its energy of dissociation.

However, this straight line is never produced into the second quadrant beyond the origin, because it is usually assumed that "we are dealing with physical systems in which the physical variables K and $\frac{1}{n^2}$ cannot take negative values." This means that the full power of Eq. (1) is arbitrarily curtailed, so that any possible errors in the fundamental theory that led to its deduction are, as it were, hidden out of sight. In the present case, since dimensionless units are used, this argument of physical units does not apply, so that both E and $\frac{1}{n^2}$ can take both negative and positive values.

The only *mathematical requirement* that restricts their signs is that *one of them must be negative and the other must be positive*. This results in the choice $(-K, +\frac{1}{n^2})$ for the line in the fourth quadrant and $(+K, -\frac{1}{n^2})$ in the second quadrant. The choice of $(-K, +\frac{1}{n^2})$ yields the straight line with slope -1 in the fourth quadrant, as well as its produced line with $(+K, -\frac{1}{n^2})$ in the second quadrant.

However, this extrapolation of the line into the second quadrant immediately brings an inherent and up to now hidden problem in the theoretical derivation of Eq. (1), and thus of Eq. (12), sharply into focus. In the second quadrant, K is positive, and $\frac{1}{n^2}$ is negative, so that it follows that the quantum number n must necessarily be a complex number, $\pm\, in$, where the $-$ sign functions in the fourth quadrant so that $\frac{1}{n^2}$ is positive, and the $+$ sign in the second quadrant where $\frac{1}{n^2}$ is negative!

This inevitable—and rather startling—conclusion clearly shows that any theory that leads to Eq. (1) is wrong, since it produces real quantum numbers. The original Balmer-type energy vs inverse quantum number-squared plots are all constrained to the fourth quadrant, leading to the belief that quantum numbers are real, and that Eq. (1) is identical with the spectroscopically obtained equation. *This is not correct, and the quantum theory must be subjected to a new and intensive scrutiny.* As said above, the origin of the problem is the way time is split off from the space coordinates in the Schrödinger and Heisenberg methodologies.

There can be no objections to the energy being negative in the fourth quadrant and positive in the second quadrant—an *arbitrary choice* was made—sometime in the past—to say that *bound states* have negative energies with respect to some imagined state where the *energy is considered to be zero*. This arbitrary choice has resulted in this case in the truncation of the straight line at $\frac{1}{n^2} = 0$, thus neglecting the other—most important—part of the information contained in the equation.

6 Conclusions

In the first part of this chapter, it was shown that the fundamental equation of atomic spectroscopy, namely, Eq. (1), very nearly but not quite correctly describes the frequencies of the experimental spectral lines of hydrogen and hydrogen-like atoms. In the second part of the chapter, it was shown that the neglect of half of the information contained in Eq. (1) by restricting its application to the negative values of E and the positive values of $\frac{1}{n^2}$ obscures the fact that the quantum numbers n should actually be imaginary numbers. Both of these analyses show that nonrelativistic quantum theory needs revision. This statement has large implications for its use in computational programs, and such programs cannot in all honesty be assumed to be based upon true quantum-mechanical principles. In addition, such computational programs deliver results which are usually interpreted in terms of a *huge, incomplete, and even inconsistent empirical "theory,"* commonly known as "*chemical bonding theory*" and its empirical and untenable concepts of energy levels and atomic and molecular orbitals.

Acknowledgements During the course of writing this chapter, I was privileged to have had many discussions with Prof. Jan Boeyens whose help and encouragement I gratefully acknowledge.

References

1. Mohr PJ, Taylor BN, Newell DB (2008) Rev Mod Phys 80:633
2. Cohen ER, Cvitas T, Frey JG, Holmström B, Kuchitsu K, Marquardt R, Mills I, Pavese E, Quack M, Stöhner J, Strauss HL, Takami M, Thor AJ (2007) Quantities, units and symbols in physical chemistry, 3rd edn. IUPAC/RSC, Cambridge, U.K.
3. Mohr PJ, Taylor BN (2000) Rev Mod Phys 72:351
4. Eides MJ, Grotch H, Shelyuto VA (2001) Phys Rep 342:63
5. Mohr PJ, Taylor BN (2002) Rev Mod Phys 77:1
6. Jentschura UD, Kotochigova S, Le Bigot E-O, Mohr PJ, Taylor BN (2005) Phys Rev Lett 95:163003
7. Jentschura U, Mohr PJ, Tan JN, Wundt BJ (2008) Phys Rev Lett 100:160404
8. Sansonetti JE, Martin WC (2005) J Phys Chem Ref Data 34:1559
9. Lide DR (ed) (1994) CRC handbook of chemistry and physics, 75th edn. CRC Press, Boca Raton
10. Moore CE (1955) Atomic energy levels, vols I and II. National Bureau of Standards, NSRDS-NBS 35 (Reissued 1971)
11. Dyson F (2006) Private communication in a letter to G. Gabrielse, July 15, 2006, as quoted by Jentschura U, Mohr PJ, Tan JN, Wundt BJ (2008) Phys Rev Lett 100:160404
12. Boeyens JCA, Schutte CJH (2012) In: Putz MV (ed) Chemical information and computational challenges in 21st century. Nova, New York
13. Putz MV (ed) (2012) Chemical information and computational challenges in 21st century. Nova, New York

Struct Bond (2013) 148: 71–91
DOI: 10.1007/978-3-642-31977-8_4
© Springer-Verlag Berlin Heidelberg 2013

Calculation of Atomic Structure

Jan C.A. Boeyens

Abstract The Thomas–Fermi and Hartree–Fock calculations of non-hydrogen atomic structure rely on complicated numerical computations without a simple visualizable physical model. A new approach, based on a spherical wave structure of the extranuclear electron density on atoms, self-similar to prominent astronomical structures, simplifies the problem by orders of magnitude. It yields a normalized density distribution which is indistinguishable from the TF function and produces radial distributions, equivalent to HF results. Extended to calculate atomic ionization radii, it yields more reliable values than SCF simulation of atomic compression. All empirical parameters used in the calculation are shown to be consistent with the spherical standing-wave model of atomic electron density.

Keywords Atomic wave model · Electron density · Golden-spiral optimization · Ionization radius · Self-similarity

Contents

J.C.A. Boeyens (✉)
Unit for Advanced Scholarship, University of Pretoria, Lynnwood Road, 0002 Pretoria, South Africa
e-mail: jan.boeyens@up.ac.za

1 Introduction

The true meaning of both quantum and relativity theories, which has been demonstrated [1] to emerge only in four-dimensional formalism, has serious implications for the three-dimensional theories of atomic and molecular structure. Nonclassical attributes of atomic matter, such as electron spin, are associated with four-dimensional hypercomplex functions, known as quaternions, and cannot be accounted for by classical three-dimensional models, which include wave mechanics as traditionally formulated. The notorious failure of quantum chemistry to model the structure of non-hydrogen atoms and molecules is a manifestation of the same problem. The awareness that atomic and molecular structures are classical three-dimensional concepts dictates the use of classical rather than four-dimensional quantum models for their characterization.

It is readily demonstrated [2] that a convergence angle of $4\pi/(2n-1)$, with integer n, generates a set of points on a golden spiral at increasing distances from the origin, reminiscent of the radii, n^2a_0 that occur in the Bohr model of the hydrogen atom. In a related study of satellites in the solar system [3], orbits were found to correspond with divergence angles that optimize the two-dimensional accretion of matter from a rotating uniform cloud, along logarithmic spirals. This consilience can hardly be accidental and accounts for the successful atomic models, independently proposed by Nagaoka and Bohr, based on the structure of Saturnian rings and planets in the solar system, respectively.

Although spectacularly successful at the time, these early atomic models failed because they followed the two-dimensional astronomical analogues, which are characterized by three-dimensional angular momentum vectors, directed along fixed rotation axes, too closely. Electronic rotation, by contrast, occurs in spherical mode, correctly described by quaternions, that give rise to the quantum-mechanical spin function. The resulting distribution of extranuclear negative charge occurs in shells around the atomic nucleus, in the form of a standing spherical wave. Although the total charge is an integral multiple of the elementary electronic charge, individual electrons cannot be distinguished within the undulating electric fluid. There are, in particular, no such things as electronic particles in this model. The elementary charge is associated with an elementary wave packet, which on further subdivision, through interaction with a positron, disperses into the vacuum.

The Bohr atomic model, which describes an electron as an orbiting particle, is well known to fail for all atoms other than hydrogen. Maxima in the optimization function should therefore not be interpreted as orbits but rather as the nodes of a spherical standing wave in line with the periodic table of the elements.

2 The Periodicity of Matter

It was first noted by Harkins [4] that the ratio $(A - Z)/A$ for the known nuclides never exceeded 0.62, later identified more accurately as the golden ratio, $\tau = 0.61803\cdots = \Phi^{-1}$ [5]. A plot of Z/N vs A of nonradioactive nuclides are shown in Fig. 1 with convergence curves as inferred by Harkins.

The region of stability is mapped more precisely by two sets of straight-line segments with inflection points at common values of A. Through these inflection points, 11 hem lines divide the field of stability such that each block contains 24 nuclides. Although there is no general agreement on half lives that define *stable* nuclides, the set of nuclides identified by different schemes never deviates seriously from the 264 selected in Fig. 1.

Replotting the data on axes of Z/N vs Z (Fig. 2) does not affect the general shape of the straight-line profiles, but the hem lines, because of their modified slopes, no longer intercept all lines of constant Z/N at the same Z.

Intersection with the lines $Z/N = 0.58,\ 0.62,\ 1.0$, and 1.04 is of special interest. The points of intersection along $Z/N = \tau$ all coincide with atomic numbers commonly interpreted to signal the completion of an electronic subshell according to the periodic table of the elements:

$$10(2p), 18(3p), 28(3d), 36(4p), 38(5s), 46(4d), 48(5s),$$
$$56(6s), 62(4f : 6/8), 70(4f), 80(5d)$$

The points at 0.58 define the periodicity implied by the wave-mechanical solution of the H electron:

$$10(2p), 18(3p), 28(3d), 36(4p), 46(4d), 52(4f : 6/8), 60(4f), 68(5p), 78(5d)$$

The relationship with the points of intersection at $N/Z = 1.0$ and at $N/Z = 1.04$ is clarified by noting how these points represent an inversion of energy levels. The points at 1.04

$$14(4f), 24(3d), 32(1s), 38(5f : 6/8), 56(4d), 62(3p), 78(6f), 88(5d), 96(3s)$$

represent the completely inverted wave-mechanical spectrum $4f < 3d < 2p < 1s\ldots$ etc. The points at 1.0 define the inverted observed periodic table of the elements.

The only known process which could invert atomic energy levels is the application of relentless pressure [6]. It becomes logical to imagine that inverted periodicity occurs where the environmental pressure on an atom approaches infinity, as in a black hole. The reciprocal situation of zero pressure correlates with the wave-mechanical assumption of nothing but the potential field of a single proton. This implies an empty universe and therefore flat space–time. By this argument,

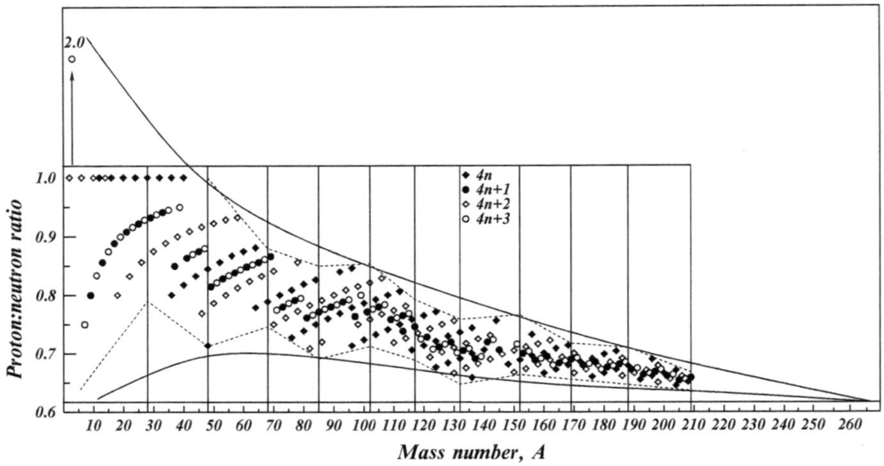

Fig. 1 Distribution of stable nuclides as a function of mass number

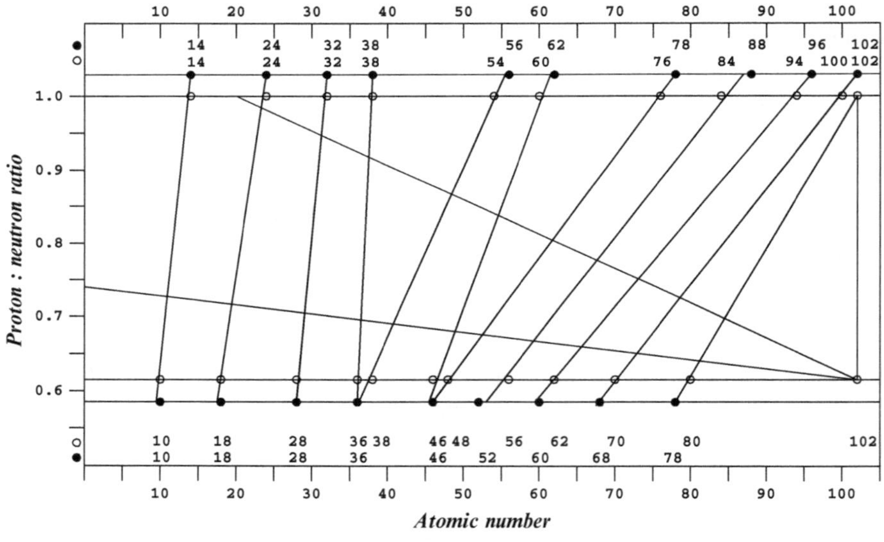

Fig. 2 Periodic functions defined by nuclide distribution

the arrangement at $N/Z = \tau$ occurs at the moderate curvature and pressure that prevails in the solar system. At $Z/N = 1$, an equilibrium between protons and neutrons is interpreted as suitable conditions for high-pressure synthesis of all nuclides from ^{4}He.

This interpretation implies a periodicity of 24 among stable nuclides, with the periodic table of the elements as a subset. It confirms that the observed periodicity

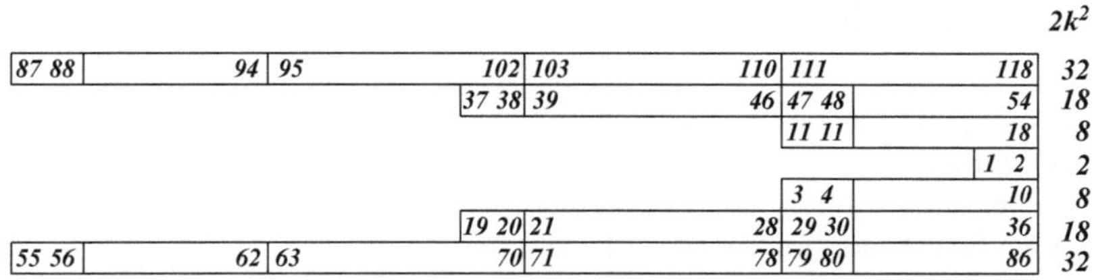

Fig. 3 Mapping of the periodic table of the elements as the reciprocal radii of the $\mathscr{F}_4$ unimodular Ford circles

is not predicted correctly by wave mechanics, but can be derived without the use of higher mathematics. The prominent role of $Z/N = \tau$ suggests a possible number-theory model of elemental periodicity.

2.1 The Periodic Table

The important Z/N ratio must by definition always be a rational fraction, and an ordered set of nuclides must therefore correspond to a Farey sequence. It is readily demonstrated [7] that a set of k-modular simple Farey fractions

$$S_k = \frac{n}{n+k},$$

not necessarily in reduced form, plotted against natural numbers, has the same appearance as Fig. 1, except for being of infinite extent. Convergence of nuclide composition from unity to $Z/N \rightarrow \tau$ implies that the limiting curves must be generated by the intersection of the infinite Farey festoons with a converging series of Fibonacci fractions. This way it could be demonstrated [6] that the points within the resulting triangle of stability represent the naturally occurring stable isotopes.

The relationship between unimodular Farey sequences and Ford circles [8] enables direct mapping of the periodic function by touching Ford circles, producing a table of the form shown in Fig. 3 [8].

The characteristic values of $Z/N = \tau$ and of 0.58 for observed and wave-mechanical periodicities are the limits of converging Fibonacci fractions around 3/5. The segmentation of the table into groups of 2 and 8 and of periods 2, 8, 18, 32 summarizes the observed periodicity as a subset of nuclide periodicity. The sublevel structure, despite formal resemblance to the wave-mechanical H solution, emerges from number theory without reference to atomic structure.

We now consider the possibility of characterizing the electronic structure of atoms as it relates to cosmic self-similarity and the periodicity of atomic matter.

3 The Golden Spiral

Casual interpretation of the local environment as three-dimensional space and universal time flow is not consistent with the known four-dimensional structure of space–time on a cosmic scale. Local Euclidean space is said to be tangent to the underlying four-dimensional curved space–time.

It has been argued [8] that the transformation from curved space–time to Euclidean tangent space is described by the golden ratio. This is not an entirely unexpected conclusion, in view of the prominence of τ in the operation of self-similar symmetries, related to equiangular logarithmic spirals.

The complex number $(a, b) = a + ib$ is represented in polar coordinates by $(r, \theta) = re^{i\theta} = r(\cos\theta + i\sin\theta)$ as in Fig. 4. Continuous rotation of the point $(a, 0)$, which transforms each point (r, θ) into $(r, \theta + t)$, has the locus $r = a$ that describes a circle. In the same way, when $(a, 0)$ is transformed by continuous *dilatation* (enlargement), the locus $\theta = 0$ of $(a, 0)$ describes the ray $(r, 0)$ for increasing $r > 0$.

Combination of these two operations represents a continuous *dilative rotation* that transforms the general point (r, θ) into $(\alpha^t r, \theta + t)$, as t changes continuously. The locus of the transform of $(a, 0)$ is the equiangular spiral

$$r = \alpha^t a, \quad \theta = t$$

i.e., $r = a\mu^\theta$. Taking the derivative

$$\begin{aligned}\frac{dr}{d\theta} &= \frac{d}{d\theta}\left(a\mu^\theta\right) \\ &= a\mu^\theta \ln\mu = r\ln\mu.\end{aligned}$$

The constant μ represents the dilatation on rotation of 1 radian. On rearrangement into

$$\frac{dr}{r} = d\ln r = \ln\mu d\theta\ ,$$

the simplest equation of the logarithmic spiral follows as $r = \mu\exp(c\theta)$, more frequently given in the classical form

$$r = ae^{\theta\cot\phi}\ ,$$

where a and ϕ are constants. It is common practice to set $a = 1$ to obtain a unit spiral.

In the special case where $\phi \simeq 73°$, $\cot\phi = \tau/2$, the spiral $r = a\exp(\theta\tau/2)$, equivalent to $r = a/\tau^{2\theta/\pi}$, corresponds almost exactly to the golden spiral [9], as constructed in a golden rectangle and shown in Fig. 5. The constant a is related to

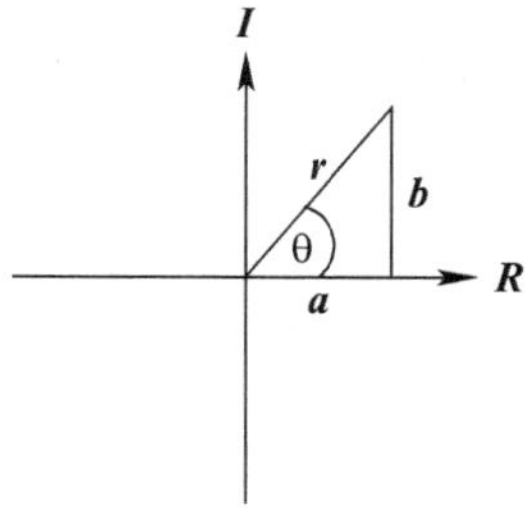

Fig. 4 The complex plane

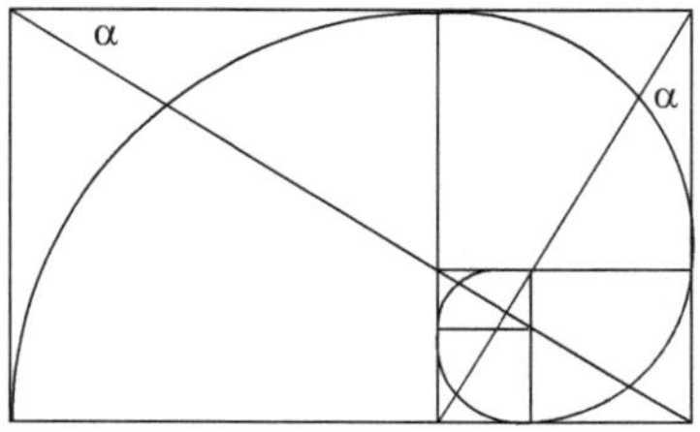

Fig. 5 The approximate golden logarithmic spiral. $\tan\alpha = \tau$. Note that $\phi \simeq 2\pi/5$ radians, emphasizing the relationship of τ to a regular pentagon: $\cos(2\pi/5) = \tau/2$

the overall dimensions of the spiral [10]. More generally, in terms of the complex number $a + \mathrm{i}b$

$$r = A\mathrm{e}^{(a/b)\theta}, \quad b \neq 0. \tag{1}$$

The distinctive property of logarithmic spirals is the constant dilatation for equal rotations. A dilative rotation of 2π transforms any point on the spiral into a *homothetic* point, which is similarly placed and directed. The spiral, said to be homothetic to itself, therefore has the property of self-similarity at all scales. We note that the origin $(r, \theta = 0)$ transforms into the homothetic points $\mathrm{e}^{n\pi\tau}$ after n rotations. We propose that this property, described by the three fundamental constants: e, π, and τ, is related to the general curvature of space–time, which is responsible for the observed cosmic self-similarity: e for growth, π for rotation, and τ for dilatation.

3.1 Self-similarity

The demonstration [1] that both Lorentz transformation and quantum spin are the direct result of quaternion rotation implies that all relativistic and quantum structures must have the same symmetry. This is the basis of cosmic self-similarity. The observation that the golden mean features in many known self-similarities confirms that τ represents a fundamental characteristic of space–time curvature. The existence of antimatter and the implied CPT[1] symmetry of space–time favors

[1]Charge conjugation-parity-time.

closed metric-free projective geometry with involution: the only topology that automatically generates the gauge invariance that links quantum mechanics to the electromagnetic field [11, 12]. This topology is consistent with constant space–time curvature, locally distorted by large gravitating masses. It seems reasonable to assume that the logarithmic spiral (1) follows the general curvature in two-dimensional projection, characteristic of stable structures and growth patterns in tangent Euclidean space. In four-dimensional space–time, the curvature is more appropriately described by a formula such as

$$\rho(x^{\mu}) = A\mathrm{e}^{(a/\sqrt{b^2+c^2+d^2})\theta} ,$$

which describes spherical rotation in quaternion notation.

The spacing of planetary orbits and of moons orbiting a planet has been shown [3] to obey simple whole number rules, not unlike the quantum rules of wave mechanics. Orbital radii, in particular, correspond to positions separated by a constant divergence angle along a golden spiral. We now demonstrate that the spacing of atomic electron shells is also related by a divergence angle, according to the same procedure.

4 Atomic Structure

Each electronic energy level is assumed to occur as a spherical shell at a distance from the nucleus, determined by a positive integer n and a minimum radius a of the innermost shell. Atomic size is wave-mechanically poorly defined. For principal quantum number n, the electronic radial distribution function has a single maximum [14] when the angular momentum quantum number l has its largest value of $n-1$. The relative density is determined by the degeneracy of $2l+1=2n-1$. Enhanced nuclear attraction and increased charge density imply that divergence angles for optimal spacing will not be constant but likely to decrease by the factor $(2n-1)$ for each quantum level. The first shell is assumed to occur at a rotation of 4π from the origin.[2] Successive divergence angles for higher levels then follow as $\phi_n = 4\pi/(2n-1)$.

Using this as a guide, the distribution of charge density, optimized by a golden spiral at a divergence angle of $4\pi/(2n-1)$, for principal quantum number n, was calculated before [2]. The simplest demonstration of such electron-density optimization is in terms of a Fibonacci spiral [8] that converges to a golden spiral with increasing Fibonacci number. Graphical derivation of orbital radii, according to this model, is shown in Fig. 6, predicting $r/a = 1, 4, 9, 16$, etc., for unit radius a, in agreement with the Bohr radii of $r_n = n^2 a_0$. The labeled points in Fig. 6 lie progressively closer to the spiral and predict to good approximation the volumes

[2]This generates spin of $\hbar/2$.

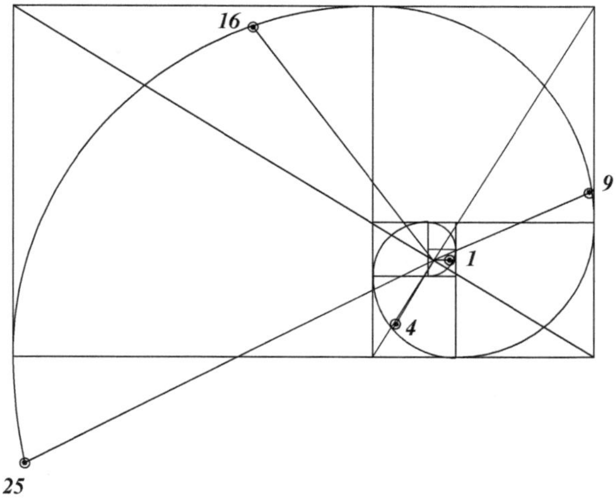

Fig. 6 A sequence of Fibonacci squares on a scale of 1:2 serves to generate the 21 cm × 13 cm golden rectangle with its inscribed spiral. Directly measurable radii of n^2a at convergence angles of $4\pi/(2n-1)$ terminate at the labeled points

Table 1 The Schrödinger table

n	Configuration	Electrons	Total
1	$1s^2$	2	2
2	$2s^2 2p^6$	8	10
3	$3s^2 3p^6 3d^{10}$	18	28
4	$4s^2 4p^6 4d^{10} 4f^{14}$	32	50
5	$5s^2 5p^6 5d^{10} 5f^{14} 5g^{18}$	50	100

and charge densities of all atoms with $Z = 1 \to 118$. On a virtually identical logarithmic spiral, $r = 1.164\exp(\theta \cdot \tau/2)$, when sampled at $\theta_n = \sum_n 4\pi/(2n-1)$, the r_n round off to n^2.

According to Schrödinger's solution for the hydrogen atom and the exclusion principle, successive energy levels can accommodate $2n^2$ electrons, as in Table 1. This arrangement cannot account for either the periodic table of the elements, the electronic structure of non-hydrogen atoms, or the predicted orbital radii.

The Thomas–Fermi (TF)[15] and Hartree–Fock (HF) schemes model atomic structure empirically, by assuming spectroscopically determined electron configurations. On the other hand, the correct form of the periodic table emerges from elementary number theory, which suggests an alternative interpretation of the Bohr radii.

Interpreted in terms of the symmetrical form of the periodic table (Fig. 3), the quantum numbers that define the radial distances of $r = n^2a$ specify the nodal surfaces of spherical waves that define the electronic shell structure. Knowing the number of electrons in each shell, the density at the crests of the spherical waves that represent periodic shells, i.e., at 1.5, 3, etc. (a), can be calculated. This density distribution, shown in Fig. 7, decreases exponentially with Z and, like the TF central-field potential, is valid for all atoms and also requires characteristic scale factors to generate the density functions for specific atoms. The Bohr-Schrödinger

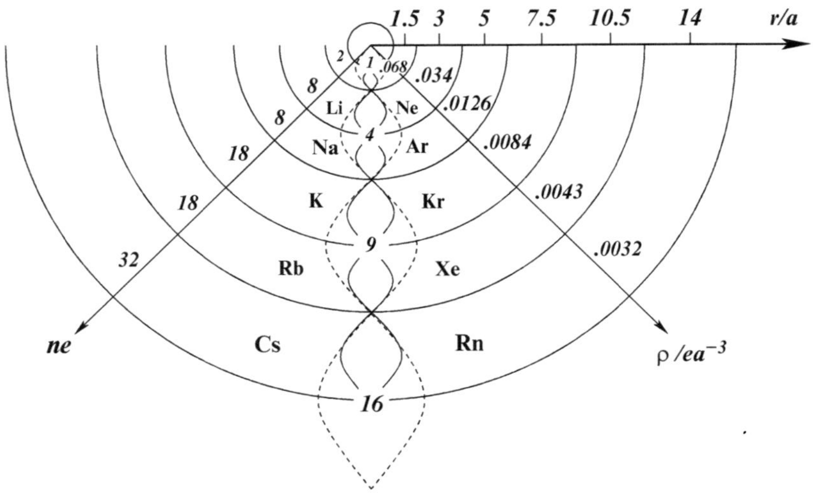

Fig. 7 Self-similar distribution of atomic electron density. For $Z = 1$, the unit of radial distance is assumed as $a = a_0$, the Bohr radius

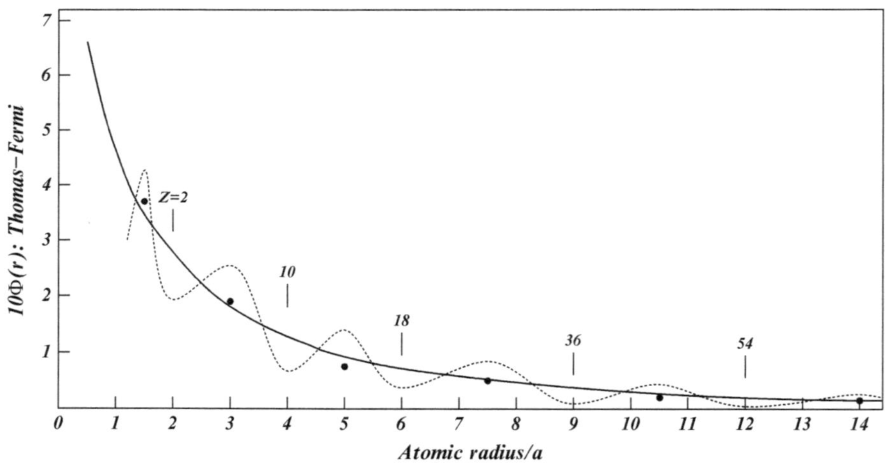

Fig. 8 Simulation of Thomas–Fermi and Hartree–Fock electron densities for unit atoms

model, represented by the stippled curve in Fig. 7, breaks down completely for non-hydrogen atoms.

The TF potential is shown as a solid curve [15] in Fig. 8, with our calculated points on the same scale (numerical factor = 55).

The curve in stippled outline is an approximate simulation of the Hartree–Fock electron density for unit atoms. A few simple assumptions allow more detailed simulation of HF results for any atom:

Table 2 Screening factors (ν) and comparison of the calculated maxima at alkali-metal wave crests ($\bar{r}_n$), scaled against Hartree–Fock [17] radial expectation values, in units of a_0

n	r'_n/a_0	Li	Na	K	Rb	Cs	Fr	Key
1	1.5	1	1	1	1	1	1	ν_i
		0.50	0.136	0.079	0.041	0.027	0.017	$\bar{r}_n$
		0.57	0.143	0.081	0.041	0.028	0.017	HF
2	3		3	2	2	2	2	
		4.05	0.818	0.316	0.162	0.109	0.069	
		3.87	0.779	0.386	0.169	0.110	0.068	
3	5			5	4	3	3	
			4.50	1.315	0.540	0.273	0.172	
			4.21	1.357	0.520	0.297	0.168	
4	7.5				7.5	5	5	
				5.07	1.620	0.682	0.431	
				5.24	1.617	0.768	0.392	
5	10.5					10.5	7.5	
					5.67	2.005	0.905	
					5.63	1.975	0.916	
6	14						14	
						6.31	2.253	
						6.31	2.168	
7	18							
							6.95	
							6.63	

1. An electron at the innermost level is not screened against attraction by the nuclear charge of $+Ze$. The one-electron radius $r' = 1.5a_0$ is thereby contracted to $r_1 = r'/Z$ [16].
2. The radii of intermediate shells contract to

$$r_i = \frac{\nu_i r'_n}{Z}$$

   where ν is a screening constant.
3. The radius of the outer shell is modeled as

$$r''_n = \frac{r'_n}{ns} .$$

Using the HF results of Mann [17] for the alkali metals as a benchmark, this screening factor could be fixed at $s = 0.37$. Appropriate values of ν_i are summarized in Table 2.

Table 3 Calculated radii $\bar{r}/a_0$ of highest electron-density maxima of p-block and alkaline earth elements, compared to the corresponding HF values [17]

μ	σ	r_n''	$\bar{r}$						
1.17		Li	Be	B	C	N	O	F	Ne
	0.59	4.05	3.27	2.04	1.75	1.49	1.27	1.09	0.93
	(HF)	3.87	2.65	2.20	1.74	1.45	1.24	1.08	0.97
1.14		Na	Mg	Al	Si	P	S	Cl	Ar
	0.8	4.50	3.62	3.16	2.77	2.43	2.13	1.87	1.64
	(HF)	4.21	3.25	3.43	2.79	2.37	2.07	1.84	1.66
1.11		K	Ca	Ga	Ge	As	Se	Br	Kr
	0.7	5.07	4.09	3.20	2.88	2.59	2.34	2.11	1.90
	(HF)	5.24	4.21	3.42	2.90	2.56	2.31	2.11	1.95
1.11		Rb	Sr	In	Sn	Sb	Te	I	Xe
	0.73	5.67	4.57	3.73	3.36	3.03	2.73	2.45	2.21
	(HF)	5.63	4.63	3.78	3.29	2.95	2.70	2.50	2.34
1.10		Cs	Ba	Tl	Pb	Bi	Po	At	Rn
	0.68	6.31	5.09	3.90	3.54	3.22	2.93	2.67	2.42
	(HF)	6.31	5.26	3.93	3.45	3.14	2.89	2.70	2.54
		Fr	Ra						
		6.95	5.60						
		6.63	5.64						

In this calculation, the calculated maxima $\bar{r}_n$ are scaled against the radial expectation values of Mann.[3] The multiplet level structure of the HF analysis is reproduced in detail. The fundamental assumption underlying this simple simulation is the indistinguishability of individual electrons in a collective, as emphasized by Schrödinger [18], Madelung [19], and Pauli [20]. This calculation has not been done for other elements, but once outer-level radial expectation values had been obtained, the procedure of Table 2 applies.

The highest radial maxima for non-alkali atoms in the same period must decrease uniformly from the alkali values r_n'', depending on the number of electrons at each sublevel. For elements of the p-block, correct values of these radial maxima are predicted as

$$\bar{r} = \frac{r_n'' \cdot \sigma}{\mu^p}$$

from characteristic values of σ and μ for a number p of electrons. The results are summarized in Table 3.

The parameters σ and $1/\mu$ are empirical estimates which describe the stepwise radial decrease of s- and p-density maxima within a periodic group. The expected periodic increase of σ with increasing period number is interrupted by atomic

[3]SCF multiplet structure is empirically based on spectroscopic results.

Table 4 Radial distances of highest maxima of d-block elements

σ	r''_n	Sc	Ti	V	Cr[a]	Mn	Fe	Co	Ni	Cu[a]	Zn
0.78	5.07	3.84	3.73	3.62		3.41	3.31	3.22	3.12		2.94
0.88					3.85					3.32	
	(HF)	3.96	3.77	3.69	3.84	3.35	3.24	3.16	3.06	3.33	2.90
	5.67	Y	Zr	Nb[a]	Mo[a]	Tc	Ru[a]	Rh[a]	Pd[b]	Ag[a]	Cd
0.76		4.18	4.06			3.72					3.21
0.85				4.28	4.16		3.92	3.80	(4.2)	3.59	
	(HF)	4.30	4.08	4.21	4.08	3.65	3.88	3.80	1.53	3.66	3.24
	6.31	Lu	Hf	Ta	W	Re	Os	Ir	Pt[a]	Au[a]	Hg
0.69		4.23	4.10	3.98	3.87	3.76	3.65	3.54			3.24
0.78									3.77	3.66	
	(HF)	4.26	4.07	3.92	3.80	3.69	3.60	3.52	3.76	3.70	3.33

[a] s^1
[b] s^0

contraction due to the first appearance of d and f levels. It will be shown that these parameters depend on the periodic variation of angular momentum and spin, and hence the exclusion principle.

The decrease of μ with periodic number refers to the relative weight of a single electron in shells of different size. At an s level, consisting of only two electrons, we estimate $\mu = 2\tau$, decreasing smoothly and converging to $\mu = 1$ with increasing electron count. Accordingly we calculate $\bar{r} = r''/2\tau$ for the alkaline earth metals, also shown in Table 3. The decrease of $\mu(s) > \mu(p) > \mu(d) > \mu(f) \to 1$ reflects the same trend.

All atoms, except for those of the p-block and of the element palladium, have an outer shell of s electrons. Atoms of the so-called d-block have a penultimate d-shell. Variation of atomic radius, within such a series with a uniform outer shell, is almost continuous. Discontinuity occurs where the number of electrons in the outer shell differs from the general s^2.

The d-block, consisting of the 3×8 transition elements and the coinage group, has radial expectation values described correctly as

$$\bar{r} = \frac{r''_n \sigma}{(1.03)^d} \quad \text{and} \quad \bar{r} = \frac{r''_n \sigma}{(1.01)^f}$$

describes inner-transition elements with an incomplete f-shell. The results are in Tables 4 and 5.

No attempt was made to calculate HF wave functions from our results, but the correspondence between ns alkali expectation values [17], $P(r)/\bar{r}$, with $\sqrt{\rho}$, measured at the wave crests, demonstrates the feasibility of such a simulation in Table 6.

Table 5 Radial distances of highest inner-transition element maxima

Cs	σ	La[a]	Ce	Pr	Nd	Pm	Sm
r_n''	0.825		5.10	5.05	5.00	4.95	4.90
6.31	0.775	4.89					
	(HF)	4.93	5.11	5.05	4.99	4.93	4.88
Eu	Gd[a]	Tb	Dy	Ho	Er	Tm	Yb
4.86		4.76	4.71	4.67	4.62	4.57	4.53
	4.56						
4.83	4.55	4.74	4.70	4.66	4.62	4.58	4.54
Fr		Ac[a]	Th[b]	Pa[a]	U[a]	Np[a]	Pu
r_n'	0.75			5.11	5.06	5.01	
6.95	0.795						5.21
	(HF)	5.23	4.98	5.11	5.05	5.00	5.18
Am	Cm[a]	Bk	Cf	Es	Fm	Md	No
	4.86						
5.15		5.05	5.00	4.95	4.91	4.86	4.81
5.13	4.86	5.05	5.01	4.97	4.93	4.90	4.86

[a] d^1
[b] d^2

Table 6 Simulated wave functions

	Li	Na	K	Rb	Cs
$\|\psi(\mathrm{HF})\|$	0.1352	0.1140	0.0840	0.0782	0.0666
$\sqrt{\rho}$	0.1844	0.1122	0.0917	0.0656	0.0566

5 Ionization Radii

The effect of applied pressure on the electronic structure of the hydrogen atom has been studied many times [21, 22] by changing the boundary condition in wave-mechanical simulation of the energy-level structure. The general effect is an increase of all energy levels with pressure, until the point is reached where the ground-state level reaches the ionization limit on compression to a radius of $r_0 = 1.835a_0$.

Such a calculation for non-hydrogen atoms was carried out numerically by a modified Hartree–Fock–Slater procedure [23]. The boundary condition for each wave function was introduced on defining a cutoff radius by the step function:

$$S = \mathrm{e}^{-(r/r_0)^p}, \quad p \gg 1 .$$

The value of p determines the sharpness of the cutoff, and it may vary from atom to atom.

A set of ionization radii with $p = 20$ was found to correspond fairly well with the characteristic atomic radii [7] that generate chemical bond dissociation energies in either point-charge or Heitler–London simulation. The value of these characteristic radii is plagued by uncertainties in thermochemical quantities and in their relationship with spectroscopic measurements.

The rationale behind this identification lies therein that the energy simulations assume uniform one-electron density within the characteristic volume, whereas an electron, decoupled from the nucleus by hydrostatic compression, is likewise confined to a sphere of radius r_0 at constant density. By exploiting this property, ionization radii were also calculated from the maxima of HFS wave functions normalized over spheres of constant density [24]. The same procedure now suggests itself for the calculation of such radii, directly from the calculated charge densities (ρ) and radial expectation values $\bar{r}$, in Fig. 7.

Normalization of the radial wave function in the ionization sphere requires

$$\frac{4\pi}{3}r_0^3\left(\frac{u^2(r)}{4\pi}\right) = 1\,.$$

The density at the radial maximum is given by $\rho = (\bar{r}/r'')[1/u^2(\bar{r})]$, and hence, in Å units,

$$r_0 = S\left(\frac{\bar{r}}{r_n''}\cdot\frac{3}{\rho}\right)^{\frac{1}{3}} a_0.$$

The scale factor increases with the size of the excluded core region. In general,

$$S = \left(1 - \frac{r_c}{r''}\right)$$

for a core of radius r_c. For group 2, $r_c = 0$ and $S = 1$. For higher groups, $r_c = 2\bar{r}_{n-2}$. For the Na group, only the $1s$ level is inaccessible, hence $S = 1 - 0.272/4.5 = 0.94$. In period 4, the appearance of a d-level between Ca and Ga results in contraction of the core, compensated for by setting $r_c = \bar{r}_{n-2}$, i.e., $S = 0.93$ for the $4p$ level. The results for representative elements of the s and p blocks are shown in Table 7.

Ionization radii calculated with the same formula and scale factors $S(4) = 0.88$, $S(5) = 0.81$, and $S(6) = 0.78$ correspond well with the values calculated by atomic compression, r_0(HFS), for the three d-series, as tabulated in Table 8.

Ionization radii for the lanthanides are in Table 9 ($S = 0.97$).

Ionization radii are of fundamental importance in chemistry. By definition, they represent the volume to which activated valence electrons are confined, and hence the quantum-potential energy of the valence state. This quantity is the same as the classical concept of electronegativity [25]. Not only is the entire theory of chemical reactivity entangled with electronegativity, but the ionization sphere also features directly in the simulation of interatomic interactions. Previous efforts to model ionization radii theoretically invariably involved some unsubstantiated assumptions. The present calculation proceeds without such assumptions, from derived extranuclear electronic arrangements.

Table 7 Ionization radii of representative elements

n	r''	ρ_n	Atom							
2	4.05	0.034	Li	Be	B	C	N	O	F	Ne
		r_0 (Å)	2.36	2.20	1.88	1.78	1.69	1.60	1.52	1.44
		r_{emp}	2.70	1.70	1.85	1.85	1.62	1.51	1.37	
		r_0(HF)	1.25	1.09	1.62	1.60	1.56	1.45	1.36	1.20
3	4.50	0.0126	Na	Mg	Al	Si	P	S	Cl	Ar
		r_0 (Å)	3.09	2.87	2.74	2.62	2.51	2.47	2.30	2.21
		r_{emp}	3.00	2.10	2.60	2.90	2.81	2.66	2.30	
		r_0(HF)	2.73	2.35	2.61	2.40	2.20	2.05	1.89	1.81
4	5.07	0.0084	K	Ca	Ga	Ge	As	Se	Br	Kr
		r_0 (Å)	3.50	3.08	3.00	2.89	2.80	2.71	2.61	2.52
		r_{emp}	3.74	2.90	2.10	3.00	2.92	2.90	2.59	
		r_0(HF)	3.74	3.26	3.29	2.94	2.62	2.40	2.28	2.12
5	5.67	0.0043	Rb	Sr	In	Sn	Sb	Te	I	Xe
		r_0 (Å)	3.81	3.54	3.31	3.20	3.09	2.98	2.88	2.75
		r_{emp}	4.10	3.43	2.90	2.80	3.40	3.30	2.92	
		r_0(HF)	4.31	3.83	3.55	3.26	3.01	2.81	2.60	2.49
6	6.31	0.00323	Cs	Ba	Tl	Pb	Bi	Po	At	Rn
		r_0 (Å)	4.03	3.75	3.43	3.32	3.22	3.12	3.03	2.93
		r_{emp}	4.30	3.74	2.80	3.10	3.19	3.50		
		r_0(HF)	4.96	4.48	3.82	3.47	3.19	3.14	3.12	3.82

Table 8 Ionization radii of d-block elements

r_0	Sc	Ti	V	Cr	Mn	Fe	Co	Ni	Cu	Zn
r_0 (Å)	3.02	2.99	2.96	3.02	2.90	2.87	2.85	2.81	2.88	2.76
r_0(HF)	3.13	3.01	2.95	2.98	2.94	2.87	2.85	2.86	2.85	2.78
	Y	Zr	Nb	Mo	Tc	Ru	Rh	Pd	Ag	Cd
r_0 (Å)	3.27	3.24	3.30	3.26	3.14	3.20	3.16	2.29	3.11	3.00
r_0(HF)	3.55	3.32	3.30	3.21	3.16	3.13	3.08	2.49	3.04	3.02
	Lu	Hf	Ta	W	Re	Os	Ir	Pt	Au	Hg
r_0 (Å)	3.54	3.50	3.47	3.44	3.40	3.37	3.34	3.41	3.38	3.24
r_0(HF)	4.24	3.83	3.57	3.42	3.38	3.37	3.23	3.16	3.14	3.12

Table 9 Ionization radii of the lanthanides

Atom			La	Ce	Pr	Nd	Pm	Sm
r_0 (Å)			4.62	4.69	4.68	4.66	4.64	4.63
r_0(HF)			4.13	4.48	4.53	4.60	4.56	4.56
Atom	Eu	Gd	Tb	Dy	Ho	Er	Tm	Yb
r_0 (Å)	4.61	4.51	4.58	4.57	4.55	4.54	4.52	4.50
r_0(HF)	4.60	4.22	4.59	4.56	4.63	4.63	4.62	4.66

6 Discussion

The calculation of atomic structure presented here assumes a wave structure of extranuclear electronic charge, distributed in a way self-similar to prominent objects, such as planetary, solar, and galactic systems, as observed in local three-dimensional space. Experience shows that all structures of this type are correctly simulated as a process of optimization based on golden spirals. For atoms, a divergence angle of $\phi = 4\pi/(2n - 1)$ predicts a spherical wave structure of period n^2a, as derived before with the Bohr model of the hydrogen atom. To be consistent with the observed symmetry of the periodic table, a well-defined shell structure that allows direct calculation of charge density is inferred. The remarkable result, for a unitary atom, is a charge distribution that scales linearly to the famous Thomas–Fermi statistical distribution, considered valid for all atoms. Whereas a TF calculation involves a sophisticated numerical procedure that depends on a critical selection of initial slope for the density curve [15], the new calculation amounts to a simple computation performed directly with a pocket calculator.

On superposition of the implied wave structure, the TF-like arrangement is transformed into a periodic curve that now resembles a distribution with the same periodic structure as a typical HF simulation of a many-electron unitary atom. To bring this result into register with actual HF models only needs a set of screening constants that regulates contraction of the density function in the field of a nuclear charge of $+Ze$. Rather than random variables, these screening constants are small numbers that reflect a variability commensurate with the periodic table.

From a chemical point of view, the outer maximum in charge density, which represents the valence density, is the most important aspect of the entire charge distribution. Having simulated the effect of compression on the valence density by HFS SCF methods myself, I am aware of the effort involved. Being able to perform, what is clearly a superior simulation, with only a pocket calculator and no further assumptions convinced me that the heavy computations of modern quantum chemistry are not needed.

The simple reason for this is now well established: quantum mechanics, like relativity, is the nonclassical theory of motion in four-dimensional space–time. All theories, formulated in three-dimensional space, which include Newtonian and wave mechanics, are to be considered classical by this criterion. Wave mechanics largely interprets elementary matter, such as electrons, as point particles, forgetting that the motion of particulate matter needs to be described by particle (Newtonian) dynamics. TF and HF simulations attempt to perform a wavelike analysis and end up with an intractable probability function. On assuming an electronic wave structure, the problem is simplified by orders of magnitude, using elementary wave mechanics. Calculations of this type are well within the ability of any chemist without expertise in higher mathematics. It has already been shown that the results reported here define a covalence function that predicts, without further assumption, interatomic distances, bond dissociation energies, and harmonic force constants of all purely covalent interactions, irrespective of bond order. In line with the philosophy that

Table 10 Relative volumes, V_e (in units of $a^3 \equiv a_0^3$ on the H-scale), of unit-electron wave packets at various atomic sublevels

P	s^2	r_s	f^{14}	d^{10}	p^6	r_p	V_T	$\sum e$	Legend
	29.3						29.3		V_{sub}
1	(15)	1.52					(15)	2	(V_e)
	τ^4						τ^4		$V_e/100$
×	Φ^2						$\Phi^{13/3}$		
	77				166		235		
2	(38)	2.09			(28)	1.88	(30)	8	
	τ^2				$\tau^{\frac{8}{3}}$		$\tau^{\frac{8}{3}}$		
×	Φ^2				Φ^2		Φ^2		
	201				436		637		
3	(100)	2.88			(73)	2.59	(80)	8	
	τ^0				$\tau^{\frac{2}{3}}$		$\tau^{\frac{1}{2}}$		
×	Φ				Φ		$\Phi^{5/2}$		
	326			1,120	705		2,151		
4	(162)	3.38		(112)	(118)	3.04	(120)	18	
	Φ			$< \Phi^{\frac{1}{3}}$	$\Phi^{\frac{1}{3}}$		$\Phi^{\frac{1}{3}}$		
×	Φ			$\Phi^{5/3}$	Φ		$\Phi^{3/2}$		
	527			2,517	1,141		4,185		
5	(262)	3.97		(252)	(190)	3.56	(233)	18	
	Φ^2			$< \Phi^2$	$\Phi^{\frac{4}{3}}$		$\Phi^{\frac{7}{4}}$		
×	Φ			$\Phi^{2/3}$	Φ		$\Phi^{7/4}$		
	853		3,668	3,552	1,846		9,919		
6	(424)	4.66	(252)	(355)	(308)	4.19	(310)	32	
	Φ^3		Φ^2	$\Phi^{\frac{8}{3}}$	$\Phi^{\frac{7}{3}}$		$\Phi^{\frac{7}{3}}$		

V_T is the calculated volume of periodic standing waves confined to spherical annuli. All (V_e) are conveniently expressed as $100\tau^m$, within rounding-off errors

molecular shape, as a classical concept, should be modeled classically, the ultimate aim is to derive the principles that underlie molecular structure and to outline a fundamental system of molecular mechanics.

Perhaps the most important result of the calculation is the way in which it confirms the number-theoretic structure of the periodic table, shown in Fig. 3, and the wave structure of the electron. From the atomic model shown in Fig. 7, the volume of each segment of Fig. 3 is calculated directly in units of a^3. The results are shown in Table 10.

All results derive from the volume of the inner shell of two electrons, $V_1 = 29.375\,a^3$. Higher two-electron sublevels (called s) and p-levels increase in size by factors of Φ^x with $x = 1$ or 2, as shown in the multiplication rows. The factors that relate the d-level volumes are fractional powers of Φ, due to the irregular occupation

Table 11 Mean empirical radii (from Table 3) of p-electrons at different levels, compared to calculated electronic radii (r_e) of the electrons, considered to be spherical waves(Table 10)

P	2	3	4	5	6
$\bar{r}_{emp}$	1.43	2.33	2.50	2.92	3.11
$r_e = \left(\frac{3V_e}{4\pi}\right)^{\frac{1}{3}}$	1.88	2.59	3.04	3.56	4.19
$\frac{3}{4}(\bar{r}/r_e)$	0.58	0.67	0.64	0.67	0.68
σ	0.59	0.8	0.7	0.73	0.68
μ	1.17	1.14	1.11	1.11	1.10
$\left(\frac{4\pi}{r_e}\right)^{\frac{1}{12}}$	1.17	1.14	1.13	1.11	1.10

All radii in units of a

numbers at these levels. The volume of a single s electron at the second level follows as $38 = 100\tau^2/a^3$. To generate the complete table from this value, we only have to assume a volume of $100\tau^{\frac{8}{3}}/a^3$ for a second period p-electron. The average one-electron volume increases steadily from $100\tau^4$ for period 1 to $100\Phi^3$ for period 6. The factor 100 is consistent with the identification of 100 natural elements [6].

We notice two further potentially meaningful trends. To good approximation, the average volume per electron (V_e) increases in even multiples of $V_1/6$, with increasing periodic level P, from $3V_1/6$ to $32V_1/3$ at $V_6 \simeq V_1 \times \Phi^{12} \simeq 100^2$. Also, a regular decrease at $P = 6$ of $V_e/100$ from $\Phi^3(\equiv \Phi^{\frac{9}{3}})$ to Φ^2 in the sequence $s > d > p > f$.

These results can only be rationalized by considering an electron as a flexible wave packet, rather than a point particle. In particular, as shown in Table 2, the volume of a given annular shell is reduced as the nuclear charge increases, which means that an electronic unit is compressed into less space. However, the effective dimensionless electronic radius of r/a remains constant as a decreases and defines the fine-structure constant as $\alpha = \sqrt{r/a}$. The dimensionless volume of the two-electron ($1s$) annulus therefore remains constant and so does the effective separation of electrons in units of a.

The effect of atomic compression is now also better understood in view of the different unit volumes at different sublevels, which correlate with orbital angular momentum. To resist compression, the large s electrons are forced into higher levels relative to p, d, and f electrons. By comparison, an f electron absorbs compressive energy by spinning more rapidly in confined space. Total inversion of relative energy toward $f < d < p < s$ must result at high pressure.

As shown in Table 11, mean electronic radii at various p-levels correlate reasonably well with the mean empirically estimated radii and the parameters σ and μ of Table 3. The estimate of $\mu = (4\pi/r_e)^{\frac{1}{12}}$ derives from spherical contraction due to increased nuclear charge, proportional to $1/r^2$, over 6 steps. These estimates only serve to show that the empirical parameters do not vary randomly.

7 Concluding Summary

Nucleogenesis in the interior of massive stellar objects yields 100 natural elements of composition $Z/(A-Z)=1$. Because of radioactive decay at reduced pressure in intergalactic space, the stability ratio converges as a function of mass number to a value of τ at $A=267=(A-Z)\Phi=Z\Phi^2$. As a result, only 81 stable elements survive in the solar system as a periodic array conditioned by τ. The observed periodicity corresponds to a Ford-circle mapping of the fourth-order unimodular Farey sequence of rational fractions.

The same periodic function results from optimization on a golden spiral with a variable convergence angle of $4\pi/(2n-1)$, which describes a spherical standing wave with nodes at n^2. Analysis of the wave structure shows that it correctly models the atomic electron distribution for all elements as a function of the golden ratio and the Bohr radius, a_0. Normalization of the wave structure into uniform spherical units simulates atomic activation, readily interpreted as the basis of electronegativity and chemical affinity.

The same model is shown to fit the electronic structure of all atoms when described in dimensionless units. The scaling symmetry observed here obeys the symmetry law of Haüy quoted by Janner [26]:

> Symmetry consists in a repeated decreasing of an object in such a way that by changing the visual distance it looks the same.

It is also known as self-similarity, a concept which is intimately related to the golden ratio, and known to operate on a cosmic scale. Our observations may therefore be rationalized by considering elementary matter as the product of large-scale space–time curvature, as described by the golden ratio. We reach the provocative conclusion that a construct, which is entirely governed by the properties of the golden ratio and number theory, predicts the electronic configuration of all atoms, without reference to any chemical know-how, as a basis of a chemical theory.

References

1. Boeyens JCA (2013) Chemistry in 4 dimensions, Struct Bond 148:25
2. Boeyens JCA (2010) Int J Mol Sci 11:4267
3. Boeyens JCA (2009) Phys Essays 22:493
4. Harkins WD (1931) Phys Rev 38:1270
5. Boeyens JCA (2003) J Radioanal Nucl Chem 259:33
6. Boeyens JCA, Levendis DC (2008) Number theory and the periodicity of matter. Springer.com
7. Boeyens JCA (2008) Chemistry from first principles. Springer.com
8. Boeyens JCA, Comba P (2013) Struct Bond 148:1
9. Coxeter HSM (1989) Introduction to geometry, 2nd edn. Wiley, New York
10. Dunlap RA (1997) The golden ratio and fibonacci numbers. World Scientific, Singapore, p 20
11. Veblen O (1933) Projektive Relativitätstheorie. Springer, Berlin. English translation in Boeyens JCA (2010) Chemical cosmology. Springer, Dordrecht, Heidelberg, London, New York
12. Veblen O, Hoffman B (1930) Phys Rev 36:810

13. Boeyens JCA (2010) Chemical cosmology. Springer, Dordrecht, Heidelberg, London, New York
14. Boeyens JCA (2003) The theories of chemistry. Elsevier, Amsterdam
15. Condon EU, Odabaşi H (1980) Atomic structure. University Press, Cambridge
16. Hinchliffe A (1999) Chemical modeling from atoms to liquids. Wiley, New York, p 200
17. Mann JB (1968) Atomic structure calculations II. Los Alamos Scientific Report, LA-3691
18. Schrödinger E (1937) Nature 140:742
19. Madelung E (1926) Z Phys 40:322
20. Pauli W, Weisskopf V (1934) Helv Phys Acta 7:709
21. Sommerfeld A, Welker H (1938) Ann Phys 424:56
22. Goldman S, Joslin C (1992) J Phys Chem 96:6021
23. Boeyens JCA (1994) J Chem Soc Faraday Trans 90:3377
24. Boeyens JCA (1978) S Afr J Chem 31:121
25. Boeyens JCA (2008) Z Naturforsch 63b:199
26. Janner A (2001) Cryst Eng 4:119

Struct Bond (2013) 148: 93–135
DOI: 10.1007/978-3-642-31977-8_5
© Springer-Verlag Berlin Heidelberg 2013

Covalent Interaction

Jan C.A. Boeyens

Abstract Reviewed in historical context, bond order emerges as a vaguely defined concept without a clear theoretical basis. As an alternative, the spherical standing-wave model of the extranuclear electronic distribution on an atom provides a simple explanation of covalent bond order as arising from the constructive and destructive interference of wave patterns. A quantitative measure derives from a number pattern that relates integer and half-integer bond orders through series of Fibonacci numbers, consistent with golden-spiral optimization. Unlike any previous definition of bond order, this approach is shown to predict covalent bond length, dissociation energy and stretching force constants for homonuclear interactions that are quantitatively correct. The analysis is supported by elementary number theory and involves atomic number and the golden ratio as the only parameters. Validity of the algorithm is demonstrated for heteronuclear interactions of any order. An exhaustive comparison of calculated dissociation energies and interatomic distance in homonuclear diatomic interaction, with experimental data from critical review, is tabulated. A more limited survey of heteronuclear interactions confirms that the numerical algorithms are generally valid. The large group of heteronuclear hydrides is of particular importance to demonstrate the utility of the method, and molecular hydrogen is treated as a special case. A simple formula that describes the mutual polarization of heteronuclear pairs of atoms, in terms of valence densities derived from a spherical-wave structure of extranuclear electronic charge, is used to calculate the dipole moments of diatomic molecules. Valence density depends on the volume of the valence sphere as determined by the atomic ionization radius, and the interatomic distance is determined by the bond order of the diatomic interaction. The results are in satisfactory agreement with literature data and should provide a basis for the calculation of more complex molecular dipole moments. The diatomic CO is treated as a special case, characteristic of all interactions traditionally identified as dative bonds.

J.C.A. Boeyens (✉)
Unit for Advanced Scholarship, University of Pretoria, Lynnwood Road, 0002 Pretoria, South Africa
e-mail: jan.boeyens@up.ac.za

Keywords Bond order · Dipole moment · Force constant · General covalence · Ionization radius · Golden ratio

Contents

1 Introduction

The simplest model of a covalent bond is based on an electrostatic point-charge simulation of overlapping spherical valence-electron charge clouds that surround monopositive atomic cores. For a homonuclear pair of atoms with radius r and internuclear distance d, the dissociation energy D is calculated from

$$D' = \varepsilon^2 \left(\frac{3}{d'} - \frac{1}{2-d'} \right) \tag{1}$$

where

$$\varepsilon = \frac{3V_0}{4\pi r^3} = 1 - \frac{3d'}{4} + \frac{(d')^3}{16}, \tag{2}$$
$$d' = d/r,$$

the common volume between the overlapping spheres

$$V_0 = \pi \left[\frac{4}{3} r^3 - r^2 d + \frac{d^3}{12} \right]$$

and $D = KD'/r$.

On equating the atomic radius to a characteristic atomic radius, r_a, a single curve of d' vs D' describes homonuclear covalent interaction, irrespective of bond order. Practical use of the formulae requires definition of a complex set of characteristic radii, which could be derived empirically [1] and was used subsequently to calculate molecular shape descriptors [2] and as the basis of a generalized Heitler–London procedure, valid for all pairwise covalent interactions [3, 4]. In all of these applications, interaction is correctly described by the dimensionless curves of Fig. 1.

For heteronuclear interaction using dimensionless distances $d' = d/R$, $R = \sqrt{r_1 r_2}$ and $r_1/r_2 = x$, the overlap formulae are

$$D' = \delta\varepsilon \left\{ \frac{1}{d'} \left[\frac{x^2 - 2x + 4}{(2-x)x} \right] - \frac{1}{(1+x)/\sqrt{x} - d'} \right\} \tag{3}$$

$$\delta\varepsilon = \left[T_3 - \frac{3T_2}{16d'} - \frac{3}{4} T_1 d' + \frac{1}{16}(d')^3 \right]^2$$

$$T_1 = \frac{1}{2}\left[x + \frac{1}{x} \right]$$

$$T_2 = x^2 + \frac{1}{x^2} - 2$$

$$T_3 = \frac{1}{2}\left[x^{3/2} + x^{-3/2} \right]$$

The general covalence curve, first calculated by the point-charge electrostatic model, has a simple geometrical construction [5] within a golden rectangle of size $2 \times 2\tau$. The limiting covalence curve AB is a semicircle centered on the extension of AD at the point ($2, 9\tau/4 = 1.39$). It is intersected at F by the homonuclear semicircle through point C and centered at coordinates (2,1). Homonuclear interactions map to this curve up to the point where it intersects AB, then follow this curve to B. All points (d', D') that characterize heteronuclear covalent bonds lie within the crescent between the two curves. The circular segment BE is centered at A.

The relationship between interatomic distance and dissociation energy of atoms in interaction has an interesting connection with the golden ratio, but is of limited use without an empirical set of characteristic atomic radii. All efforts to derive such radii from atomic properties have been unsuccessful for the simple reason that these radii are not free-atom properties. However, the search has resulted in the identification of a useful set of free-atom radii, characteristic of the atomic valence state [6]. Whereas the interaction radii r_a describe the relationship between interatomic distance and dissociation energy, free-atom valence radii predict these quantities separately, but related via bond order, which is defined precisely in terms of interfering spherical electron waves. In this chapter, we show how these predictions agree quantitatively with available spectroscopic, crystallographic and thermodynamic data. All observed bond lengths, dissociation energies and stretching force constants are taken from the tables in HCP [7].

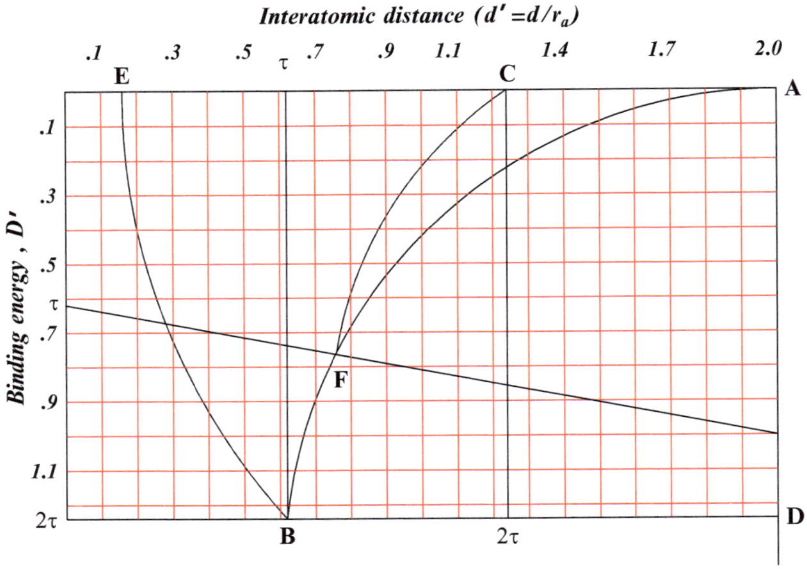

Fig. 1 Covalence curves in dimensionless units. Homonuclear interactions are described by the curve BFC and heteronuclear interactions map into the crescent CFA

2 The Bond-Order Concept

The bond-order concept is developed from the notion of multiple bonding, which was formulated empirically towards the end of the nineteenth century, to explain the composition of organic compounds. The basic rule of thumb, proposed by Kekulé [8], was to assign valences of 1–4 to H, O, N and C, respectively. In order to rationalize the observed composition of ethylene, acetylene and benzene, it was necessary to postulate the formation of double, triple and $1\frac{1}{2}$-order carbon–carbon bonds in these compounds.

This scheme could be extended in a natural way to O, N, S, P, Cl, etc. To bring the scheme into line with the stereochemical ideas of Le Bel and van't Hoff, valences came to be associated with *affinity centers*, arranged tetrahedrally in the surface of spherical atoms. The formation of single to triple bonds was explained on this basis by Meyer [9, 10] as the touching of atomic spheres in different mode as shown in Fig. 2.

Whereas the atomic centers remain at the same distance from each other, the affinity centers move progressively further apart as the bond order increases, predicting weaker interaction.

Based on Sommerfeld's atomic model [11] of elliptic orbits, directed towards the corners of a cube, a number of chemists, including Kossel, Lewis, Langmuir and Bury, developed an electronic theory to account for atomic structure and valency at

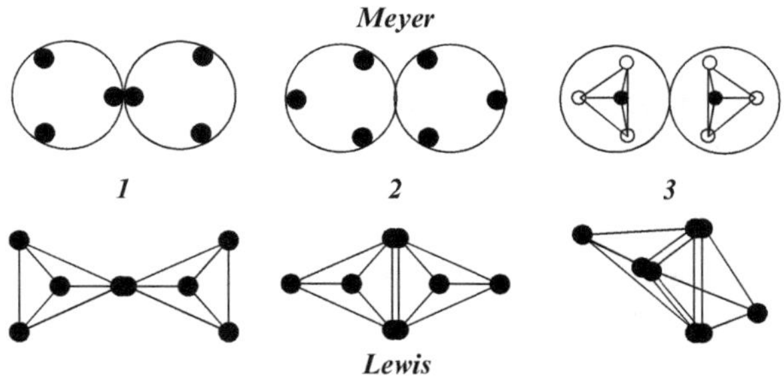

Fig. 2 Bond orders distinguished in terms of affinity centers (*top*) and electron pairs (*bottom*)

the same time. To account for the periodic table of the elements, Rydberg's formula for the atomic number of an inert gas

$$N = 2(1^2 + 2^2 + 2^2 + 3^2 + 3^2 + 4^2 + \cdots)$$

had to be reconciled with spectroscopic evidence, supported by Sommerfeld's atomic model, which predicted the number of electrons per shell as equal to the square of the principal quantum number, i.e. $n^2, n = 1, 2, 3 \ldots$. Instead of assuming that by increasing the number of electrons, atomic shells would become occupied to saturation in the order of increasing n, it was postulated that the completion of the next octet (at the corners of a cube) takes priority over saturation of the shell.

In the case of covalent interaction, the octet is of primary importance, next to the role of electron pairs, implied by the factor 2 in Rydberg's formula, from which twofold symmetry in the electronic configuration of an atom was inferred. It is important to note that the classification of electrons into s, p, d, f subsets, characterized by four quantum numbers in terms of Pauli's exclusion principle, which states that no two electrons in one atom can have all four quantum numbers identical, antedates the development of wave mechanics.

In line with van't Hoff's stereochemistry and the orientation of elliptic orbits, covalent bonds could be represented by tetrahedra that touch in apical, edgewise and facial mode, involving one, two or three electron pairs in an interaction, also shown in Fig. 2. This theory predicts increased bond strength with increasing bond order, but fails to account quantitatively for observed internuclear distances. For example, this model predicts the interatomic distances in methane and acetylene in the ratio of 3:1.

Occurrence of the stereo isomers, known as maleic and fumaric acids, has been interpreted for a long time as evidence of a barrier to rotation around a double bond. It is of interest to note that this steric rigidity is consistent with the orientation of Victor Meyer's affinity centers and with the Lewis model of electron pairs.

The definition of bond order as the number of electron pairs shared between two atoms is still widely accepted today, but the geometry of interaction has been adapted to the theory of orbital hybridization, to be considered next.

2.1 Orbital Hybridization

Schrödinger's rationalization of atomic spectra and integral quantum numbers, in terms of a wave model, introduced an improved mathematical description of the electronic configuration of atoms, derived empirically before. Unfortunately, the theory was not developed to its full potential because of reluctance to abandon the classical concept of subatomic particles.

The coming of wave mechanics, which should have been hailed as final vindication of the proposed wave nature of electrons, already surmised and soon to be demonstrated experimentally at the time, was successfully resisted as inadequate to account for cloud-chamber trajectories and the Compton effect. Although both objections are spurious, they had such authority in support that an illogical watered-down reinterpretation of Schrödinger's model gained universal acceptance. The consequences for theoretical chemistry have been disastrous.

In summary, Schrödinger managed to solve a differential equation that describes the motion of an electron in the central field of a proton, as in the hydrogen atom, in wave formalism. By separating the radial and angular components of the wave function, three quantum numbers, essentially equivalent to those of Sommerfeld, were obtained without further assumption. The three quantum numbers, which obey the exclusion principle, are conveniently summarized as follows:

$$n = 1, 2, 3, \ldots$$

$$l = 0, 1, \ldots, (n-1)$$

$$m_l = -l, \ldots, +l$$

in conjunction with the empirically added spin quantum number, $m_s = \pm\frac{1}{2}$. In physical interpretation, the principal quantum number, n, specifies the eigenvalues of the electronic energy, whereas l and m_l specify the eigenvalues of orbital angular momentum and its value in a magnetic field, respectively.

Although the Schrödinger solution is demonstrably superior to the Sommerfeld model, it lacks the pictorial appeal of the Lewis tetrahedral model. Still, there was the general belief articulated by Linus Pauling [12] that

> ... if quantum theory had been developed by the chemist rather than the spectroscopist it is probable that the tetrahedral orbitals described below would play the fundamental role in the theory, in place of the *s* and *p* orbitals.

The chemist has not succeeded in doing this. Pauling himself proposed the scheme of orbital hybridization as a quantum theory of covalent interaction. Despite its uncritical acceptance for many years, this approach is shown by elementary reasoning to be in direct conflict with the fundamentals of quantum theory.

It starts with a degenerate set of orbital angular momentum vectors with quantum numbers $l = 1$, $m_l = -1, 0, 1$, which in cartesian coordinates may be formulated as

$$p^{-1} = \frac{x - \mathrm{i}y}{r}; \quad p^0 = \frac{z}{r}; \quad p^1 = \frac{x + \mathrm{i}y}{r}.$$

The use of complex quantities is avoided by making the linear combinations:

$$\frac{1}{2}(p^{-1} + p^1) = \frac{x}{r}; \quad \frac{1}{2}(p^1 - p^{-1}) = \frac{\mathrm{i}y}{r}.$$

The overall result is clearly equivalent to the new set:

$$p^{-1} = \frac{z - \mathrm{i}y}{r}; \quad p^0 = \frac{x}{r}; \quad p^1 = \frac{z + \mathrm{i}y}{r},$$

which represents a simple rotation of the coordinate axes.

Pauling, however, preferred a different interpretation by defining

$$p_y = \frac{1}{2\mathrm{i}}\left(p^1 - p^{-1}\right) = \frac{y}{r},$$

in order to generate a degenerate set of real p-"orbitals",

$$p_x = \frac{x}{r}; \quad p_y = \frac{y}{r}; \quad p_z = \frac{z}{r},$$

directed along the cartesian axes. This procedure destroys the complex entanglement of the nonclassical variables, demanded by quantum theory, to produce three orthogonal functions with $m_l = 0$, in violation of the exclusion principle, which is not required in classical systems. Efforts to associate electron spin with real orbitals are therefore meaningless.

No amount of handwaving can circumvent this conclusion. The elaborate procedure whereby these orbitals are incorporated in further "hybridization" to define the combinations sp^3, sp^2 and sp to simulate tetrahedral, trigonal and linear sets of orbitals is likewise without quantum-mechanical meaning [13]. At best, it amounts to a classical reconstruction of these geometries. In short, the well-known procedure to define bond order and steric rigidity in terms of overlapping σ and π orbitals is meaningless, representing no more than the Lewis model, in more dignified jargon. The fanciful notion of π-overlap cannot explain why a triple bond should have no barrier to rotation rather than twice the rigidity of a double bond.

2.2 *Bond Order in Molecular Mechanics*

The only successful simulation of molecular conformation, based on classical concepts, has become known as molecular mechanics. It relies on the ideas of chemical bonds free of strain and computerized minimization of the strain energy generated by distortion of the strain-free interactions in a molecule. In this application, it is necessary to stipulate strain-free bond lengths for bonds of different order. Although such parameters can in many cases be derived empirically, a more fundamental theoretical prediction would, for obvious reasons, be preferred. Efforts to derive suitable parameters by the methods of quantum chemistry have been futile.

Important progress became possible on noting a simple relationship between interatomic distances in bonds of different order. The rationale behind the observation comes from the simple model of a covalent bond seen as the situation of equilibrium between the electrostatic attraction of a pair of valence electrons to the nuclei and the internuclear repulsion. In the formation of higher-order bonds, the role of those valence electrons in excess of bonding pairs may be seen as screening the internuclear repulsion [15]. The logic behind this interpretation is supported by the observation that, given the details of any bond, addition of a universal screening constant to the interaction transforms the single bond into a bond of specified higher order, irrespective of the atoms involved.

In practice, calculations have been performed in two different ways. Given the bond length, dissociation energy and stretching force constant characteristic of the single bond, the interaction is described by a Morse function. If this function is modified by addition of a term that represents screening of the internuclear repulsion, the relevant Morse curve of the higher-order bond is obtained [16]. Alternatively, the potential-energy curve, calculated by the Heitler–London method [3], is modified in the same way, using the same screening factors, to simulate higher bond orders.

Heitler–London simulation of general covalence depends on a set of characteristic atomic radii, assumed to describe a single electron in the valence state. Such radii were obtained empirically [17], in the first instance, by point-charge simulation of covalent interaction [1]. A more satisfactory derivation of atomic radii was discovered in the simulated compression of atoms in Hartree–Fock calculations, resulting in ionization at a characteristic compression, closely related to the empirical radii [18].

These ionization radii, which have been shown [19] to underpin the electronegativity concept, have recently been derived by an extremely simple and more reliable simulation of atomic structure as a standing electronic wave packet [6]. This simulation, which is free of the errors of approximation that affect the HF simulation of small atoms, has produced a more reliable set of ionization radii, suitable for direct prediction of interatomic distance in general pairwise interaction within bonds of any order. The procedure is outlined in the next section.

2.3 Bond Order from Ionization Radii

Comparison of the interatomic distances (d) reported for homonuclear covalent interactions, commonly considered to be first order, revealed a remarkable

Table 1 Calculated interatomic distance for low-order homonuclear interactions

	C	N	O	F
r_0 (Å)	1.78	1.69	1.60	1.52
$d(X-X)$	1.545	1.467	1.491	1.417
Expt.	1.54	1.47	1.48	1.41
	Si	P	S	Cl
r_0 (Å)	2.62	2.51	2.47	2.30
$d(X-X)$	2.274	2.179	2.144	1.996
Expt.	2.32	2.21	2.05	1.99
	Ge	As	Se	Br
r_0 (Å)	2.89	2.80	2.71	2.61
$d(X-X)$	2.51	2.43	2.35	2.27
Expt.	2.41	2.44	2.32	2.28
	Sn	Sb	Te	I
r_0/Å	3.19	3.09	2.98	2.88
$d(X-X)$	2.77	2.68	2.59	2.68
Expt.			2.59	2.67

relationship with the corresponding ionization radii, r_0. Using data from HCP [7], the large majority of bond lengths defined in dimensionless units as $d' = d/r_0$ had $d' = 0.868$, with little variation. A few notable exceptions occurred for F–F, O–O and I–I with $d' = 0.932$. Supporting evidence for typical interactions is shown in Table 1.

On repeating the exercise for traditional second-order bonds, a similar result of $d' = 0.764$ is obtained. For third-order bonds, with an admittedly smaller sample, one calculates $d' = 0.680$. Using the observed bond length of benzene, one finds $d' \simeq 0.786$ for the bond of assumed $1\frac{1}{2}$ order. Extending the search to homonuclear transition-metal diatomic molecules where high-order bonds are common and to diatomic alkali metals with assumed bond orders of zero, a complete set of dimensionless bond lengths, in good agreement with experiment, was established for all orders.

If we define zero bond order to occur at $d = r_0$, i.e. $d' = 1$, an interesting variation with bond order, from unity to the golden ratio, $\tau = 0.61803\ldots$ is inferred. This variation is reminiscent of the convergence of the ratio $Z/(A-Z)$, of protons to neutrons in stable nuclides, that leads to the generalized periodic function of atomic matter [20]. Using this as a cue, the variation of bond length with bond order can be specified as a power series in τ. In fact, any power $n > 6$, with integers j_b as coefficients defines bond order b, by $d' = j_b\tau^n$, as shown in Table 2.

The coefficients for given bond order increase like a Fibonacci series with increasing n. This is immediately obvious for the coefficients of bond orders 4 and 0, which correspond, in both cases, to the familiar Lucas numbers. This correspondence is interpreted to define a closed, and hence periodic, system,

Table 2 Any power of the golden ratio, τ^n with covariant j_b, tabulated in *bold script*, as coefficients, predicts the ratio $d/r_0 = d' = j_b\tau^n$, which determines the dimensionless bond length of order b

Order, b	τ^4	τ^5	τ^6	τ^7	τ^8	τ^9	τ^{10}	τ^{11}	τ^{12}	τ^{13}
4	**4**	**7**	**11**	**18**	**29**	**47**	**76**	**123**	**199**	**322**
Δj			1	1	2	3	5	8	13	21
$3\frac{1}{2}$			**12**	**19**	**31**	**50**	**81**	**131**	**212**	**343**
Δj				1	1	2	3	5	8	13
3				**20**	**32**	**52**	**84**	**136**	**220**	**356**
Δj			1	1	2	3	5	8	13	21
$2\frac{1}{2}$	**5**	**8**	**13**	**21**	**34**	**55**	**89**	**144**	**233**	**377**
Δj			1	1	2	3	5	8	13	21
2			**14**	**22**	**36**	**58**	**94**	**152**	**246**	**398**
Δj			1	1	2	3	5	8	13	21
$1\frac{1}{2}$			**15**	**23**	**38**	**61**	**99**	**160**	**259**	**419**
Δj		1	1	2	3	5	8	13	21	34
1		**9**	**16**	**25**	**41**	**66**	**107**	**173**	**280**	**453**
Δj		1	1	2	3	5	8	13	21	34
$\frac{1}{2}$		**10**	**17**	**27**	**44**	**71**	**115**	**186**	**301**	**487**
Δj		1	1	2	3	5	8	13	21	34
0	**7**	**11**	**18**	**29**	**47**	**76**	**123**	**199**	**322**	**521**

The different coefficients for consecutive bond orders are related by the Fibonacci numbers Δj

Table 3 Definition of d' derived from r_0 for different bond orders

Order	d'	ε	$n/40$	$n(2)$	$n(3)$	$n(4)$	$n(5)$	$n(6)$
0	1.000	0.3	12	7	10	10	11	11
$\frac{1}{2}$	0.935	0.35	14	6	8	9	10	10.5
1	0.869	0.4	16	5	7	8	9.5	10
$1\frac{1}{2}$	0.804	0.425	17	4.5	6.5	7.5	9	9.5
2	0.764	0.45	18	4	6	7	8.5	9
$2\frac{1}{2}$	0.724	0.475	19					8.5
3	0.683	0.50	20	3			8	8
$3\frac{1}{2}$	0.658	0.525	21				7.5	7.5
4	0.618	0.55	22				7	7

The relative overlap volume ε appears quantized in units of $n/40$. The columns on the *right* are discussed in the next section

consistent with the assumed spherical-wave structure of a valence electron. By noting how Δj_b for any pair of consecutive bond orders also defines a Fibonacci series with increasing n, the appropriate coefficients for any power can be specified directly without calculation. The empirically derived bond-order scale factors d' are then seen to be integral multiples of τ^n. It could be of special significance in the analysis of aromatic interactions to note that $d'(0) = \tau^0$, $d'(1\frac{1}{2}) = 1/(2\tau)$ and $d'(4) = \tau$. The sufficiently converged values of $j_b\tau^{13}$ are shown in Table 3.

With reference to the point-charge simulation of covalent interaction, we note that the common volume between two overlapping spheres of radius r, with centers at a distance d apart, is calculated as

$$V_0 = \pi\left[\frac{4}{3}r^3 - r^2 d + \frac{d^3}{12}\right],$$

or, in dimensionless units of $d' = d/r$, the relative overlap volume

$$\varepsilon = \frac{3V_0}{4\pi r^3} = 1 - \frac{3d'}{4} + \frac{(d')^3}{16}. \qquad (2)$$

This quantity, also listed in Table 3, is seen to assume quantized values of $n/40$ for the common bond orders.

2.4 Dissociation Energy and Bond Order

It is generally accepted that there is some inverse relationship between covalent bond length (d) and dissociation energy (D). The point-charge model of covalent interaction defines this relationship in terms of a smooth curve (Fig. 1) which represents all homonuclear diatomic interactions on expressing distance and energy in special dimensionless units, defined by $d' = d/r_0$, $D' = Dr_0/K$, where K is a dimensional constant. For D in kJ mol^{-1} or eV, respectively, $K = 1,389$ or 14.35. Having shown that the bond order-related linearity $d = j_b r_0 \tau^n$ is generally obeyed, we infer fixed values of d' for all bonds of order b.

The observed relationship between the common volume, defined by overlapping charge spheres, and bond order, shown in Table 3, suggests a direct relationship between bond order and dissociation energy. Noting the connection with spherical volume, we look for a dependence of the type

$$\frac{D_x r_0}{K} = D' \propto r_0^3$$

and find that first-order homonuclear interactions for p-block elements obey the rule

$$D' = r_0^3 \tau^n, \quad \text{i.e.} \quad D_x = K r_0^2 \tau^n. \qquad (4)$$

The values of n, which produce dissociation energies, D_c, to match experimental data D_x, correlate positively with bond orders derived from interatomic distances. Some results are shown in Table 4.

Results for some higher-order bonds in the format A(n) : $D_c(D_x)$ include

C(4):642(600); C(3.16):964(966); N(3):937(945); O(4):519(498)
S(6):472(425); As(7):374(382); Se(7):351(331); C(4.5):505(479 in biphenyl)

Table 4 Dissociation energies (kJ mol^{-1}) and exponents n for lowest-order homonuclear observed interactions in the s and p blocks

	Li	Be			B	C	N	O	F
r_0 (Å)	2.36	2.20			1.88	1.78	1.69	1.60	1.52
n	9	10			6	5	6	6	6
D_c	102	54			273	397	221	198	179
D_x	110	59			290	377	252	214	159
	Na	Mg			Al	Si	P	S	Cl
r_0	3.09	2.87			2.74	2.62	2.51	2.47	2.30
n	11	12			9	7	6	7	7
D_c	66	35			138	328	487	292	253
D_x	75	11			133	310	485	286	243
	K	Ca	Cu	Zn	Ga	Ge	As	Se	Br
r_0	3.50	3.08	2.88	3.11	3.00	2.89	2.80	2.71	2.61
n	12	13	8.5	13	10	8	8	8	8
D_c	53	25	192	22	100	247	232	217	202
D_x	57	~17	201	22(6)	<106	264	181	223	194
	Rb	Sr	Ag	Cd	In	Sn	Sb	Te	I
r_0	3.81	3.54	3.11	3.00	3.31	3.19	3.09	2.98	2.88
n	13	14	9	15	11	9	8	8	9
D_c	38	21	163	7	76	186	282	263	152
D_x	49	16	159	7	82	187	302	258	153
	Cs	Ba	Au	Hg	Tl	Pb	Bi		
r_0	4.03	3.75	3.38	3.24	3.43	3.32	3.22		
n	13		9	15	12	11	9		
D_c	43		209	10	51	77	190		
D_x	44		226	8	63	87	197		

Within a periodic family, interactions with common n have equal bond orders. Calculated n, for first-order interaction, increases stepwise from $n = 5, 6$ for second period elements, to $n = 10$ for period 6. We find $n = 5$ for C and $n = 6$ for N, O, F, previously identified to form $\frac{1}{2}$ order bonds. For second- and third-order interactions, within a given period, the appropriate exponents are $n_2 = n_1 - 1$ and $n_3 = n_1 - 2$. This rule would restrict golden exponents to integers and half integers. All interactions with $d > r_0$ are traditionally described as nonbonded.

Bond-by-bond data to compare calculated parameters with experiment are presented in Sects. 3 and 4.

2.5 Stretching Force Constants

The relationship between bond order and dimensionless interatomic distance is represented by three linear functions over the intervals $b = (0, 1\frac{1}{2})$, $(1\frac{1}{2}, 3)$ and (3,4) as shown in Fig. 2.

The different slopes correspond to the differential change of d with respect to b and hence describe the response of bond order to increasing d, commonly defined as a stretching force constant:

$$\frac{1}{2}k_r = \frac{\Delta D'}{(\Delta d')^2},$$

In molecular mechanics, the usual practice is to specify energies in units of $\mathrm{kJ\,mol^{-1}}$, interatomic distance in Å and k_r in $\mathrm{N\,cm^{-1}}$ ($\equiv\mathrm{mdyne\,Å^{-1}}$). In these units,

$$\begin{aligned} k_r &= \frac{2\Delta D' \times K \times 10^{-2}}{6.2r_0(r_0d')^2} = \frac{K\Delta D'}{301(\Delta d')^2 r_0^3} \\ &= \frac{4.615\Delta D'}{(\Delta d')^2 r_0^3}\,\mathrm{N\,cm^{-1}} \end{aligned} \tag{5}$$

For interactions of known bond order, the quantities $\Delta d'$ for a stretch to lower order follow directly from Table 1, provided it occurs in a region of uniform slope. Energy differences are in general proportional to $\Delta D = \tau^n - \tau^{n+1} = \tau^{n+2} = \tau^+$. As bond order is not an absolute measure, only relative slopes can be stipulated. Also, whereas d' may change continuously, bond order changes in discrete steps, determined by the interference of standing waves. We therefore assume that equation (5) correctly describes a hypothetical linear response for a stretch from bond order four ($d' = 0.6$) to zero, with a slope of 0.1 measured in Fig. 3. For any intermediate stretch of different slope σ, a multiplicative scale factor that represents an effective slope of $s = \sigma/0.1$ is added to Eq. (5).

The procedure is illustrated by calculating the stretching force constant of diatomic C_2, considered of bond order $2\frac{1}{2}$, such that

$$k_r = \frac{4.615 \times (\tau^{5.5} = 0.0709)}{(0.080 \times 1.78)^2} \times 0.8 = 12.59\,\mathrm{N\,cm^{-1}}$$

The scale factor of 0.8 represents the slope. Note that the stretch is considered to operate between bond orders $2\frac{1}{2}$ and $1\frac{1}{2}$, i.e. $n = 3.5 \leftarrow 4.5$, $\Delta D \propto \tau^{5.5}$. This result is in good agreement with the experimental $12.16\,\mathrm{N\,cm^{-1}}$.

As a further test, the force constant of third-order N_2 ($\tau^+ = \tau^{3+2}$) follows as

$$k_r = \frac{4.615 \times 0.0902}{(0.081 \times 1.69)^2} = 22.2 \quad \mathrm{N\,cm^{-1}},$$

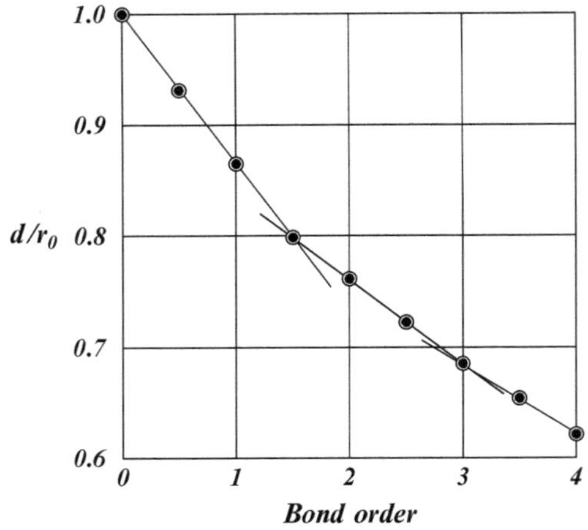

Fig. 3 Variation of interatomic distance (d') with bond order

in exact agreement with experiment.

Calculation of k_r for $C\overset{2}{—}C$ is complicated by the change of slope between orders 2 and 1. This is readily compensated for by graphical recalculation of $\Delta d' = 0.093$ and mean slope of 0.93:

$$k_r = \frac{4.615 \times 0.0554}{(0.093 \times 1.78)^2} \times 0.93 = 8.68\,\mathrm{N\,cm^{-1}}$$

in good agreement with the value of $8.43\,\mathrm{N\,cm^{-1}}$ measured for $CCl_2 = CH_2$.

For $C\overset{1}{—}C$, the energy difference is obtained from $\tau^5 - \tau^{6.5} = 0.0464$ and

$$k_r = \frac{4.615 \times 0.0464}{(0.131 \times 1.78)^2} \times 1.3 = 5.12\,\mathrm{N\,cm^{-1}}.$$

The force constant calculated from the vibrational frequency ω_e of ethane is $k_r = 4.50\,\mathrm{N\,cm^{-1}}$.

For F_2, the energy difference $\tau^5 - \tau^6 = \tau^7$ occurs against a calculated force constant $k_r = 5.2$, compared to the experimental $4.7\,\mathrm{N\,cm^{-1}}$.

The calculated force constant for O_2 is obtained from $\Delta d' = 0.093$, $n^+ = 6$ as $k_r = 10.79$, compared to the observed $11.77\,\mathrm{N\,cm^{-1}}$.

Extension of the calculation to higher periods relies on the wave model of atomic electron density. Changes in bond order are interpreted as stepwise changes in the pattern of overlap between the electronic wave structures of interacting atoms. The effect on interatomic distance depends on the wavelength of the interfering waves, which in turn depends on atomic volume as elaborated in Sect. 2.6. In the second shell of 8 electrons, $\Delta d' = 0.1306$ corresponds to unit change in bond order. At the next level with an additional 8 electrons, a stretch of $\Delta d' = 0.0653$ suffices for relaxation to lower order.[1] With some empirical guidance, we arrive at the unimodular sequence of factors 1, $\frac{1}{2}$, $\frac{2}{5}$, $\frac{1}{3}$, $\frac{1}{4}$ to effect a change of bond order at successive electronic levels. The scheme is demonstrated by the calculated force constants for diatomic halogens:

$$Cl_2 : \Delta d' = 0.0653, n^+ = 9, k_r = 2.7 \times 1.3 = 3.51 \quad (3.23\,\mathrm{N\,cm^{-1}})$$

$$Br_2 : \Delta d' = 0.0522, n^+ = 10, k_r = 2.62 \quad (2.46\,\mathrm{N\,cm^{-1}})$$

$$I_2 : \Delta d' = 0.0435, n^+ = 11, k_r = 1.91 \quad (1.72\,\mathrm{N\,cm^{-1}})$$

Also on the second period

$$Si_2 : \Delta d' = 0.0653, n^+ = 9, k_r = 2.71 \quad (2.15\,\mathrm{N\,cm^{-1}})$$

and the second-order molecules

$$P_2 : \Delta d' = 0.093/2 = 0.0465, n^+ = 8, k_r = 5.78 \quad (5.56\,\mathrm{N\,cm^{-1}})$$

$$S_2 : \Delta d' = 0.0465, n^+ = 8, k_r = 5.96 \quad (\mathrm{HCP{:}}\ 4.96\,\mathrm{N\,cm^{-1}})$$

The reported value [7] for S_2 is obviously in error.

Quadruple dimetal interactions provide an interesting test:

$$Cr_2 : \Delta d' = 0.0522, n^+ = 9, k_r = 2.45$$

$$Mo_2 : \Delta d' = 0.0375, n^+ = 9, k_r = 4.08$$

$$Re_2 : \Delta d' = 0.0375, n^+ = 9, k_r = 3.75$$

$$W_2 : \Delta d' = 0.0375, n^+ = 9, k_r = 3.66$$

The effective slope in the high-order region is close to unity. These results, in $\mathrm{N\,cm^{-1}}$, are in exact agreement with empirical molecular-mechanics simulations of these force constants [21–24].

2.6 *Wave Model of Bond Order*

The idea of covalent bond order is of special importance in the present instance, but not in its traditional form as the number of electron pairs shared between two atoms.

The familiar electron-pair exposition of bond order is that C, in the valence state, has four unpaired electrons after promotion of an *s* electron. In the case of N, there is

---

[1] A bond is stretched by external forces, such as steric interactions, only until it flips spontaneously into the wave pattern that stabilizes lower bond order.

no room in the p-subshell to allow such promotion. However, the positive ion N^+, like C, has four unpaired electrons and may form four single bonds as in $NH_4{}^+$. Oxygen has two unpaired electrons and is restricted to form either two single bonds or one double bond. The scheme works, but fails to account for the paramagnetism of dioxygen with known second-order bond strength. The experimental facts are consistent with neither of the following:

It is generally believed that the bonding in O_2 is correctly described in terms of molecular orbitals. However, this method, as traditionally formulated, argues the involvement of p_x, p_y and p_z orbitals, which have no physical meaning, and the bond order of dioxygen remains a mystery. This is the most glaring, but by no means the only, failure of the electron-pair definition of bond order.

The very idea of electron pairs, which presupposes point particles of charge, does not feature in the model of covalence proposed here. A plurality of electrons at the same energy level is considered instead as a single multiply charged standing wave with quantized orbital angular momentum and spin. We propose that bond order of diatomic domains within molecules is quantized in a similar way, and we look for a numerical sequence to account for the empirical regularity of Table 3. Prominence of the golden ratio in this formulation suggests a sequence based on Fibonacci fractions and/or the golden logarithmic spiral. This would render it self-similar to structures in the solar system, the periodic table of the elements and the electronic configuration of atoms.

As a first trial, we consider a series of Fibonacci fractions in the range 1/1 to 3/5 to simulate bond orders between 0 and 4 ($d' = 1, \tau$). The unimodular sequence that converges to 1, i.e.

$\frac{1}{1}$	$\frac{7}{8}$	$\frac{6}{7}$	$\frac{5}{6}$	$\frac{4}{5}$	$\frac{3}{4}$	$\frac{2}{3}$	$\frac{3}{5}$
1	0.875	0.857	0.833	0.8	0.75	0.667	0.6
0	1			$1\frac{1}{2}$	2	3	4

is immediately seen to reflect many features of the bond-order function.

An even better simulation, based on this sequence, is obtained by selecting suitable terms from a single Farey sequence, which means that all terms have the same denominator. This way we derive from $\mathscr{F}_{15}$ the sequence

$\frac{1}{1}$	$\frac{14}{15}$	$\frac{13}{15}$	$\frac{4}{5}$	$\frac{11}{15}$	$\frac{2}{3}$	$\frac{3}{5}$	$\left(\frac{n}{15}, n = 9 - 15\right)$
1	0.933	0.867	0.8	0.733	0.67	0.60	
0	$\frac{1}{2}$	1	$1\frac{1}{2}$	2	3	4	

Fig. 4 Simulation of integer bond orders on a golden logarithmic spiral

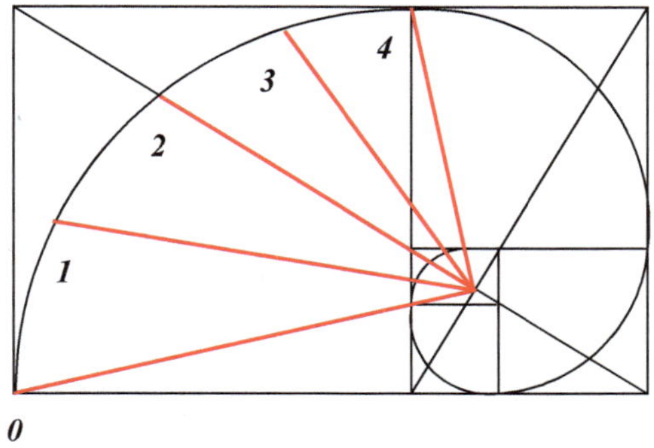

Starting from the next Fibonacci fraction, 5/8, ($\mathscr{F}_{16}$)

$\frac{1}{1}$	$\frac{15}{16}$	$\frac{7}{8}$	$\frac{13}{16}$	$\frac{3}{4}$	$\frac{11}{16}$	$\frac{5}{8}$	$\left(\frac{n}{16}, n = 10 - 16\right)$
1	0.938	0.875	0.81	0.75	0.688	0.625	
0	$\frac{1}{2}$	1	$1\frac{1}{2}$	2	3	4	

The next Fibonacci fraction predicts ($\mathscr{F}_{13}$)

$\frac{1}{1}$	$\frac{12}{13}$	$\frac{11}{13}$	$\frac{10}{13}$	$\frac{9}{13}$	$\frac{8}{13}$	$\left(\frac{n}{13}, n = 8 - 13\right)$
1	0.923	0.846	0.769	0.69	0.615	
0	$\frac{1}{2}$	1	2	3	4	

Starting from still higher Fibonacci fractions, the same pattern persists, but gaps appear in the sequence of quantum numbers. The infinite sequence between 1 and τ is inferred to have the exact bond-order sequence, with large quantum numbers, embedded within it.

The most convincing simulation of bond order is from a golden logarithmic spiral with convergence angle of $\pi/8$, for integer orders, or $\pi/16$ to include half-integer orders, shown in Fig. 4. Bond orders 0 and 4 are separated by a right angle.

This simulation confirms the results of Table 3 in detail. As a matter of interest, all of these bond orders are approximated in $\mathscr{F}_{34}$, which relates to the Fibonacci fraction 21/34, with a few gaps:

$\frac{1}{1}$	$\frac{15}{17}$	$\frac{29}{34}$	$\frac{27}{34}$	$\frac{13}{17}$	$\frac{25}{34}$	$\frac{23}{34}$	$\frac{11}{17}$	$\frac{21}{34}$
1	0.941	0.853	0.794	0.764	0.735	0.676	0.647	0.618
0	$\frac{1}{2}$	1	$1\frac{1}{2}$	2	$2\frac{1}{2}$	3	$3\frac{1}{2}$	4

The apparent quantization of bond order corresponds to the numerators in Farey sequences that converge to the golden ratio. As the limiting Fibonacci fraction $n/(n+1) \rightarrow \tau$ approaches the golden ratio, the values of quantized bond order, predicted by the Farey sequence $\mathscr{F}_{n+1}$, approach the simulation of Fig. 4.

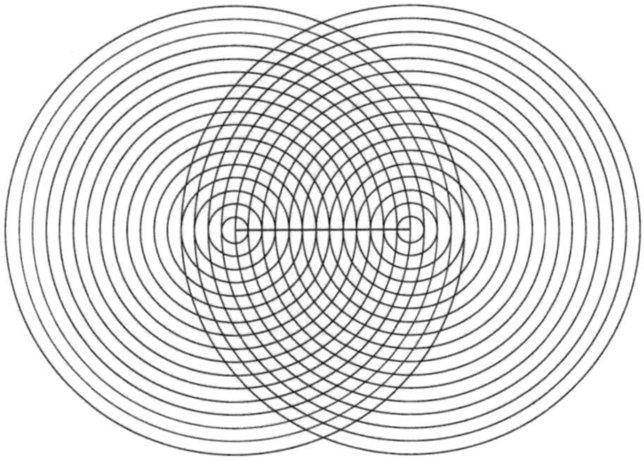

Fig. 5 Wave structure that defines second-order homonuclear interaction

Table 5 Comparison of different estimates of bond order

n	33	32	31	30		29	28		27	26	25		24	23		22	21
1,000n/34	971	941	912	882		853	823		794	764	735		706	676		647	τ
Mean		941			868			809				721			662		
Table 2		935			869			804		764		724		683	658		τ
BO		$\frac{1}{2}$			1			$1\frac{1}{2}$		2		$2\frac{1}{2}$		3	$3\frac{1}{2}$		4

The simulation described here gives new meaning to the bond-order concept. It now emerges as the quantization of interatomic distance within molecular diatomic domains.

The quantization results from constructive interference between the valence-electron waves of interacting atoms of opposite spin. The interference pattern for second-order homonuclear interaction, $d' = 26/34 \equiv 13/17$, is shown in Fig. 5. Only even numbered wave crests are shown. Bond order changes in regular steps as the dimensionless interatomic distance, d', changes in steps of 1/34, which define the wavelength of the spherical waves. For C and H with respective values of $r_0 = 1.78$ and 0.97 Å, wavelengths of $\lambda(\mathrm{C}) = 1.78/34 = 5.24 \times 10^{-12}$ m and $\lambda(\mathrm{H}) = 2.86 \times 10^{-12}$ m are predicted. This result is in winsome agreement with the Compton wavelength of an electron at 2.43×10^{-12} m.

Gaps in the 21/34 sequence of bond orders suggest the possibility of intermediate bond orders at 31/34, 28/34 and 24/34. This conclusion is supported by the comparison (Table 5) with the bond orders predicted by Table 2. One finds that the calculations converge and confirm earlier empirical conclusions:

- Between bond orders 0 and 1, there is a broad, poorly resolved, shallow minimum of weak interactions, collectively assigned to bond order $\frac{1}{2}$.
- Bond orders 2 and 4 are well resolved.

- Bond orders 1, $1\frac{1}{2}$, $2\frac{1}{2}$ are less sharply defined, but within clear limits.
- Bond orders 3 and $3\frac{1}{2}$ are poorly resolved.
- The $\mathscr{F}_{34}$ sequence specifies an absolute measure of bond order, but the Table 2 values are more convenient in practice.

The idea of bond lengths that may vary around special integer bond orders was first proposed on empirical grounds [25].

The findings reported here provide new evidence for the unity of micro- and macrophysics and refute the perception of separate quantum and classical domains. The known universe exists as a four-dimensional space–time manifold but is observed in local projection as three-dimensional Euclidean tangent space that evolves in universal time. The observable world, at either micro- or macroscale, can be described in either four-dimensional (nonclassical) or in classical three-dimensional detail. The descriptive model may change, but the reality stays the same. This realization is at the root of self-similarity between large and small. The symmetry operator, which reflects the topology of space–time, is the golden logarithmic spiral.

Cosmic self-similarity has been documented and discussed many times with reference to atomic nuclei, atomic structure, the periodicity of matter, covalence, molecular conformation [26], biological structures, planetary and solar systems [27], spiral galaxies and galactic clusters [28]. The prominent role of the golden ratio in all cases can only mean that it must be a topological feature of space–time structure.

The simplest and most beautiful illustration of golden-ratio self-similarity must surely be the quantization of bond order, shown in Figs. 3 and 4. It is so intimately entangled with golden symmetry and gives such a precise definition of the otherwise elusive bond-order concept that the possibility of this being mere coincidence is zero. Small wonder that the great Johannes Kepler referred to the "divine proportion" which "served as idea to the Creator when He introduced the creation of likeness out of likeness, which also continues indefinitely".

In the same spirit, the construction of Fig. 5, which has nothing to do with chemistry, may well be used as a starting point from which to derive a theory of covalent interaction, atomic structure, elemental periodicity and molecular shape. Working backwards through the concepts developed before [5, 6, 26] a complete framework of chemistry may be developed as a theme in pure number theory.

The natural limit to bond order, which occurs at $d' = \tau$, $D' = 2\tau$, as a result of the wave nature of electrons, is inferred to reflect the topological property that limits electron density in space–time to a natural maximum. It is the molecular counterpart of the property that limits the total charge density at the electronic energy levels on an atom. The empirical rule, known as Pauli's exclusion principle, is formulated in terms of either spin pairing or the antisymmetry of four-dimensional wave functions, which amount to the same thing. As currently understood, it appears as an emergent property of matter waves, with its origin in space–time topology.

3 Homonuclear Interaction

For ease of reference, we tabulate atomic ionization radius (r_0), bond order (b) and exponents (n) of lowest-order observed homonuclear interactions in Table 6.

Based on the parameters in Table 6, calculated parameters for homonuclear low-order interactions and diatomic molecules are compared with experimental data in Table 7. Calculation involves the characteristic atomic radii, r_0, and the scale factors based on the golden ratio:

$$d_c = d_b' r_0$$

$$D_c = K r_0^2 \tau^n \, .$$

Having tested these formulae exhaustively against all spectroscopic and crystallographic data in HCP [7], it is found that they work without serious exception. The bond order relationship between d and D holds generally, but a few cases deserve special mention.

The relationship is most sensitive for the small atoms of period 2. Based on dissociation energies, the atoms B to F have the exponent $n = 6$, except for C with $n = 5$. Judging by interatomic distance, however, B and C appear to be of first order whereas O, N and F have $b = \frac{1}{2}$. We interpret this trend in terms of increasing spectator electron density in the relatively small valence shells of these atoms. The dissociation energies seem to indicate a gradual decrease in bond order from C to F of the form: $1, < 1, > \frac{1}{2}, \frac{1}{2}$. Higher-order bonds of these atoms will be shown to have golden exponents $n = 6 - b$.

The exponent $n = 8.5$ indicates $b = \frac{1}{2}$ for Cu, but the observed interatomic distance of 2.22 Å in diatomic Cu_2 is typically first order, in line with the variability exposed in Table 5.

Metals of the second and third transition series are well known to be characterized by multiple dimetal interactions of orders 3, $3\frac{1}{2}$, and 4 [29]. The large reported errors in the measured diatomic dissociation energies for some of these metals are interpreted as due to spectroscopic activation, producing equilibrium mixtures of compounds of poorly resolved bond order. It is noted that in all such cases, an average over two bond orders reproduces the experimental data rather well.

Diatomic W_2 provides an interesting demonstration of an interaction which is prevented from reaching bond order 4 ($n = 7$) by the exclusion principle that restricts maximum D' to $2\tau = 1.236$. Since $D_4' = 1.4 > 2\tau n$ is restricted to the minimum of 7.3.

Observed interatomic distances for diatomic transition-element interactions are estimates of the fraction, $d = 0.78\delta$ of nearest-neighbor approaches in the metals [5] and may be considerably in error in the present context, especially for the second transition series. Apart from first-order La_2 and Ce_2, with $D_x = 245 \pm 30\,\mathrm{kJ\,mol^{-1}}$, homonuclear diatomics have weak interactions with an average $D_x = 70 \pm 40\,\mathrm{kJ\,mol^{-1}}$ in agreement with our estimates. Multiple bond orders, in general, are characterized by stepwise reduction of the first-order golden exponent, such that

Table 6 Parameters r_0 (Å), b and n

	Li	Be	B	C	N	O	F	Ne		
r_0	2.36	2.20	1.88	1.78	1.69	1.60	1.52	1.44		
b	< 0		1	1	$\frac{1}{2}$	$\frac{1}{2}$	$\frac{1}{2}$			
n	9	10	6	5	6	6	6	14		
	Na	Mg	Al	Si	P	S	Cl	Ar	K	Ca
r_0	3.09	2.87	2.74	2.62	2.51	2.47	2.30	2.21	3.50	3.09
b	0		$\frac{1}{2}$	1	2	1	1		< 0	
n	11	12	9	7	6	7	7	15	12	13
	Sc	Ti	V	Cr	Mn	Fe	Co	Ni		
r_0	3.02	2.99	2.96	3.02	2.90	2.87	2.85	2.81		
b	1	1	2	1	$\frac{1}{2}$	1	1	2		
n	9	9	8	9	10	9	9	8		
	Cu	Zn	Ga	Ge	As	Se	Br	Kr	Rb	Sr
r_0	2.88	2.76	3.00	2.89	2.80	2.71	2.61	2.52	3.81	3.54
b	1		0	1	1	1	1			
n	8.5	13	10	8	8	8	8	15	13	14
	Y	Zr	Nb	Mo	Tc	Ru	Rh	Pd		
r_0	3.27	3.24	3.30	3.26	3.14	3.20	3.16	~ 2.5		
b	1	3	4	$3\frac{1}{2}/4$	$3/3\frac{1}{2}$	$3/3\frac{1}{2}$	2	2		
n	9.5	8	7	7.5/7	8/7.5	8/7.5	8.5	8.5		
	Ag	Cd	In	Sn	Sb	Te	I	Xe		
r_0	3.11	3.00	3.31	3.19	3.09	2.98	2.88	2.75		
b	1		0	1	2	2	1			
n	9	15	11	9	8	8	9	15		
	Cs	Ba	La	Ce	Pr	Nd	Pm	Sm		
r_0	4.03	3.75	4.62	4.49	4.68	4.66	4.64	4.63		
b			1	1	$\frac{1}{2}$	0	0	0		
n	13		10	10	11	12				
	Eu	Gd	Tb	Dy	Ho	Er	Tm	Yb		
r_0	4.61	4.51	4.58	4.57	4.55	4.54	4.52	4.50		
b			$\frac{1}{2}$		0					
n	14		11	12.5	12	12.5	13	15		
	Lu	Hf	Ta	W	Re	Os	Ir	Pt		
r_0	3.54	3.50	3.47	3.44	3.40	3.37	3.34	3.41		
b	1	2	$3/3\frac{1}{2}$	< 4	$3/3\frac{1}{2}$	$3\frac{1}{2}$	3	$2\frac{1}{2}/3$		
n	10	9	8/7.5	7.3	8/7.5	7.5	8	8.5/8		
	Au	Hg	Tl	Pb	Bi	Po	At	Rn		
r_0	3.38	3.24	3.43	3.32	3.22	3.12	3.03	2.93		
b	2	< 0	$\frac{1}{2}$	$\frac{1}{2}$	2					
n	9	15	12	11	9	7				

Table 7 Homonuclear bond length/pm and matching D (kJ mol^{-1})

	Li	Be	B	C	N	O	F	Ne		
d_c	*	*	163	155	147	149	142			
d_x	267		159	154	147	148	141			
D_c	102	54	273	397	221	198	179	3		
D_x	110	59	290	377	252	214	159	4		
	Na	Mg	Al	Si	P	S	Cl	Ar	K	Ca
d_c	309	*	256	227	218	214	200		*	*
d_x	308	389	247	232	221	205	199		392	
D_c	66	35	138	328	487	292	253	5	53	25
D_x	75	11	133	310	485	286	243	5	57	17
	Sc	Ti	V	Cr	Mn	Fe	Co	Ni		
d_c	262	260	226	262	271	249	247	215		
d_x	279	252	228	243	300	216	218	217		
D_c	167	163	259	167	95	151	148	233		
D_x	163	118	269	152	81	118	127	204		
	Cu	Zn	Ga	Ge	As	Se	Br	Kr	Rb	Sr
d_c	250	*	300	251	243	235	227		*	*
d_x	222			241	244	232	228		432	
D_c	192	22	100	247	232	217	202	6	38	21
D_x	201	22	106	264	181	223	194	5	49	16
	Y	Zr	Nb	Mo	Tc	Ru	Rh	Pd		
d_c	284	221	203	208	211	215	241	191		
d_x	309	250	210	221	235	231	245	239		
D_c	152	311	520	454	332	344	232	145		
D_x	159	298	513	436	330	331	236	136		
	Ag	Cd	In	Sn	Sb	Te	I	Xe		
d_c	270	*	*	277	268	259	288			
d_x	251					259	268			
D_c	163	7	76	186	282	263	152	7		
D_x	159	7	82	187	302	258	153	7		
	Cs	Ba	La	Ce	Pr	Nd	Pm	Sm		
d_c	*		401	408	438	466				
d_x	470									
D_c	43		240	247	152	94				
D_x	44		247	242	130	84				
	Eu	Gd	Tb	Dy	Ho	Er	Tm	Yb		
d_c	*		428	*	455	*	*			
D_c	35		145	70	89	69	54	20		
D_x	33		132	71	86	75	54	21		
±	17		25	29	30	29	17	17		

(continued)

Table 7 (continued)

	Lu	Hf	Ta	W	Re	Os	Ir	Pt	
d_c	307	267	233	225	228	220	227	240	
d_x	299	246	223	214	238	233	236	248	
D_c	141	224	405	490	387	427	330	307	
D_x	142		390	486	386	415	361	307	
±	33		96	96	96	77	68	2	
	Au	Hg	Tl	Pb	Bi	Po	At	Rn	*
d_c	258	*	320	288	246				$>r_0$
d_x	247				266				
D_c	209	10	51	77	190				
D_x	226	8	63	87	197	187			

Table 8 Comparison of interatomic distance (Å) and dissociation energy (kJ mol^{-1}) for second-order interaction

	C	O	P	S	As	Se	Sb	Te
d_c	1.36	1.22	1.92	1.89	2.14	2.07	2.36	2.28
d_x	1.34	1.21	1.89	1.89	2.10	2.15		2.56
D_c	642	519	484	472	374	351	282	263
D_x	600	498	485	425	382	331	302	258

$n_{1+i} = n_1 - i$. Some observed second-order homonuclear interactions in the p-block are collated in Table 8.

Improved estimates of $d = 2.18$ and 2.58 for Se_2 and Te_2, respectively, indicate bond orders of $1\frac{1}{2}$ and 1 rather than 2. The only authentic homonuclear third-order interaction in the p-block occurs for diatomic N_2. It has $d_c = 1.15$ Å, $d_x = 1.10$; $D_c = 937\,\text{kJ mol}^{-1}$, $D_x = 945$. What is commonly considered to be a triple dicarbon interaction is approximated by $n = 3.16$, as restricted by the 2τ limit. This way $d_c = 1.22$ Å, $d_x = 1.21$; $D_c = 964\,\text{kJ mol}^{-1}$, $D_x = 966$. The non-existence of a third-order diphosphorous interaction is explained directly by noting that it would imply $D' \gg 2\tau$.

The dicarbon interaction of order $1\frac{1}{2}$, predicted to have the graphitic $d = 1.40$ Å and $D = 505\,\text{kJ mol}^{-1}$, has been measured in biphenyl with $D_x = 479\,\text{kJ mol}^{-1}$.

4 Heteronuclear Interaction

Bond-order analysis of heteronuclear covalent interactions is considerably more complicated, but feasible in principle. An obvious assumption, $R_0 = \sqrt{r_0(1) \cdot r_0(2)}$, predicts correct values for $d = d_b' R_0$ and suggests $D = pKR_0^2\tau^n$. The parameter $p = \chi(1)/\chi(2)$ is taken as the ratio of electronegativities $\chi(i) \propto r_0^2(i)$ [19] to compensate for polarization effects. Hence,

$$D_c = Kr_0^3(1)\tau^n/r_0(2)\,,$$

with $r_0(1) > r_0(2)$.

The formulae work surprisingly well when tested against a myriad of examples. A self-consistent set of effective exponents n and bond orders for the p-block is shown in Table 9. Readers are urged to verify the numbers against the large volume of data on heteronuclear diatomic molecules [7], some of which are collated in Table 10. The reported $D_x = 96\,\text{kJ mol}^{-1}$ for AsSe appears suspect and has been ignored.

The most interesting molecule in the group is CO, reported to have the strongest covalent bond at 1,076 kJ mol^{-1}. Classically, it is formulated as :C$\overset{\leftarrow}{\equiv}$O: , with a so-called dative triple bond. Our formulation is in agreement with such a special structure as a regular third-order interaction would exceed the limit of $D' = 2\tau$ imposed by the exclusion principle. To put the situation into perspective, it is noted that on ionization into CO^+, the interatomic distance decreases from 1.128 to 1.115 Å, which seems to imply an increase in bond strength. However, the observed dissociation energy also decreases in the process from 1,076 to 806 kJ mol^{-1}. Referred to Fig. 1, this CO interaction lies outside the covalent crescent, CBA, which also indicates an additional factor, such as an ionic contribution to the total interaction, at work.

What seems to be happening is that, when prevented from establishing third-order interaction, a rearrangement of the combined valence density occurs in such a way that a more efficient lower-order interference pattern is promoted. Such a rearrangement exists in a modification of the atomic valence spheres. An outward flow of electron density causes a decrease in characteristic radius, and vice versa. A decrease of $r_0(\text{O})\rightarrow$ 1.36 Å, balanced by an increase of $r_0(\text{C})\rightarrow$ 1.784 Å, is found to promote the formation of $2\frac{1}{2}$-order interaction at $R_0 = \sqrt{1.36 \times 1.784} = 1.56$ Å, to match the observed $d = 1.56 \times 0.724 = 1.128$ Å and dissociation energy $D_x = 1,389\tau^{3.5} \times 1.784^3/1.36 = 1,076\,\text{kJ mol}^{-1}$, as observed.

Ionization, $CO\rightarrow CO^+ + e$, implies decrease of both characteristic radii. At $r_0(\text{O})$=1.44 and $r_0(\text{C})$=1.65 Å the $2\frac{1}{2}$-order interaction occurs at $R_0 = \sqrt{1.44 \times 1.65} = 1.54$ Å to match the observed $d = 1.54 \times 0.724 = 1.115$ Å and dissociation energy $D_c = K\tau^{3.5} \times 1.65^3/1.44 = 804\,\text{kJ mol}^{-1}$, as observed. This polarization also explains the observed dipole moment of CO.

The same bonding pattern repeats for all groups 14–16 diatomic molecules that map outside the covalent region of Fig. 1.

4.1 Hydrides

The hydrides constitute the largest group of heteronuclear covalent interactions. Calculated results are in Tables 11 and 12.

Table 9 Bond orders and exponents for diatomic interactions between representative elements

	F	Cl	Br	I	O	S	Se	Te		N	P	As	Sb	Bi
Li	1½	1	1	½	1				B	2½				
	6.5	6	7	7.5	7.5	7				5.5	7.5			
Na	1	1	½	½	½				C	3	2			
	8.5	8	8	8	9.5					4	6.5			
K	½	½	½	½					N		2½	2		
	9	8.5	8.5	8.5	10						6.5	7.5	7.5	
Rb	½	½	½	½	½				O	3	2	2		
	9.5	9	9	9	10.5					4	6.5		8.5	9.5
Cs	½	½	½	½	½				F	1½	1½		1	
	10	9.5	9.5	9.5	11					5.5	7.5	7.5		9.5
Be	1				2½	2			Al					
	5.5	6	7.5		5	7				8	9	8.5		
Mg	1	1			1½				Si	2				
	8	7.5	7.5	8	8	8.5				7.5	7			
Ca	1	1		½	1½	1½			P					
	8.5	8	8.5	8.5	9	8.5						7	8	9
Sr	1	1	1	½	1½				S					
	9	8.5	8.5	9	9.5	9				7	6.5		8	8.5
Ba	1	½	½	½	1½	1½			Cl	1½	1½			
	9.5	9	9	9	9	9				7	7	7		8.5
B	2	1	1	½	3	2			Ga					
	5	6	7.5	8	4	6	7	8.5			8.5	8.5	9	9.5
Al	1½	1	1	1	2	1½			As					
	7	6.5	7	7	7.5	7.5	7	8					8	
Ga	1½	1	1	1	1½				Se					
	8	7.5	7.5	7.5	8.5			8			7			8.5
In	1	1	1	1					Br	1				
	9	8.5	8.5	8.5	9.5	9	9	9		8.5	7			9
Tl	1	1	1	1					In					
	9.5	9	9	9	10.5						9.5	9.5	9.5	9.5
C	2	1½			2½	2½	2		Sn					
	5	6.5	8	9	3.5	6	7							9
Si	1	1			2½	2	2		Sb					
	7	7	7.5	8	5.5	6	6	7						8.5

(continued)

Table 9 (continued)

	F	Cl	Br	I	O	S	Se	Te		N	P	As	Sb	Bi
Ge	1	1	1		2	2	2	$1\frac{1}{2}$	Te					
	8	7.5	7.5	8	7	6.5	6.5	7			8		8	8.5
Sn	1	1			$1\frac{1}{2}$	$1\frac{1}{2}$	$1\frac{1}{2}$	$1\frac{1}{2}$	Tl					
	8.5	8.5	8	9	8.5	7.5	8	8			9.5		10	10.5
Pb	$\frac{1}{2}$				$1\frac{1}{2}$	$1\frac{1}{2}$	$1\frac{1}{2}$	$1\frac{1}{2}$	Pb					
	9.5	9	9	9.5	9	8	8.5	8.5						10

Table 10 Calculated and experimental dissociation energies (kJ mol^{-1}) of heteronuclear first-order interactions

	C	N	F	Si	P	Cl	Ge	As	Br	I
C		*	*				248			
		5.5	5				248			
N	329		*							
	329		5.5							
O	347		208			225		318		
	351		220			206		302		
F	465	313				*			*	*
	514	<349				6			6	6.5
Si	298		565							
	302		567							
S	318		293			247				
	294		286			256				
Cl	326	167	270	374	329		310	282	*	*
	336	155	261	349	328		340	294	7.5	8
As	226		427							
	231		466							
Se	259		304							
	244		286							
Br	296	118	305	329	267	226	274	249		
	277	118	280	315	265	219	277	244		
Sn	205					327				
	227					319				
Sb	186									
	197									
I	246		266	270	174	196	194	198	*	
	214		<272	235	185	211	181	181	8.5	
Xe			153							
			130							

Golden exponents *n* are marked by asterisks

Table 11 Bond order and interatomic distance in H–X interactions

	Li	Be	B	C	N	O	F
R_0 (Å)	1.59	1.47	1.35	1.31	1.28	1.25	1.21
b	0	0	$\frac{1}{2}$	1	$1\frac{1}{2}$	$1\frac{1}{2}$	2
d (Å)		1.47	1.26	1.14	1.01	0.98	0.92
Obs	1.60	1.34	1.23	1.12	1.04	0.97	0.92
	Na	Mg	Al	Si	P	S	Cl
R_0	1.73	1.68	1.63	1.59	1.56	1.55	1.49
n	< 0	0	0	$\frac{1}{2}$	$\frac{1}{2}$	1	1
d		1.68	1.63	1.48	1.45	1.34	1.30
Obs	1.89	1.73	1.65	1.52	1.42	1.34	1.27
	K	Ca	Ga	Ge	As	Se	Br
R_0	1.84	1.74	1.71	1.67	1.65	1.62	1.59
n	< 0	< 0	0	$\frac{1}{2}$	$\frac{1}{2}$	$\frac{1}{2}$	1
d			1.71	1.56	1.49	1.51	1.38
Obs	2.44	2.00	1.67	1.59	1.51	1.47	1.41
	Rb	Sr	In	Sn	Sb	Te	I
R_0	1.92	1.86	1.79	1.76	1.73	1.70	1.67
n	< 0	< 0	0	0	0	0	$\frac{1}{2}$
d			1.79	1.76	1.73	1.70	1.56
Obs	2.37	2.15	1.84	1.78	1.70		1.61
	Cs	Ba	Tl	Pb	Bi	Au	Hg
R_0	1.98	1.92	1.82	1.79	1.77	1.81	1.77
n	< 0	< 0	0	0	0	1	0
d			1.82	1.79	1.77	1.57	1.77
Obs	2.49	2.23	1.86	1.80	1.75	1.52	1.74
	Cr	Mn	Ni	Cu	Zn	Ag	Cd
R_0	1.71	1.68	1.65	1.67	1.64	1.74	1.71
n	0	0	1	1	0	$\frac{1}{2}$	0
d	1.71	1.68	1.43	1.45	1.64	1.62	1.71
Obs	1.67	1.73	1.47	1.46	1.59	1.62	1.76

4.2 Stretching Force Constants

The calculation of harmonic force constants of covalent bonds has been shown to derive from ionization radii by an equally simple procedure. Following the rule defined as

$$D' = r_0^3\tau^n, \quad \text{i.e.} \quad D_x = Kr_0^2\tau^n$$

it follows directly that for homonuclear interactions

$$k_r = \frac{4.615\tau^+ s}{(\Delta d' \cdot r_0)^2},$$

Table 12 Calculated and experimental dissociation energies (kJ mol^{-1}) of H–X interactions

	Li	Be	B	C	N	O	F	
n	9	8	7	6.5	6.5	5.5	4.5	
D_c	248	201	327	353	303	416	577	
D_x	238	200	340	338	<339	430	570	
	Na	Mg	Al	Si	P	S	Cl	Cu
n	11.5	11.5	9.5	9.5	9	8.5	7.5	10
D_c	167	136	303	265	298	360	472	277
D_x	186	126	288	293	297	354	431	255
	K	Ca	Ga	Ge	As	Se	Br	Ag
n	12	11.5	10.5	10	10	9.5	9	10.5
D_c	190	167	247	280	255	294	336	276
D_x	181	223	276	263	274	313	366	244
	Rb	Sr	In	Sn	Sb	Te	I	Au
n	13	12.5	11	11	10.5	10.5	10	10.5
D_c	150	152	260	232	270	243	277	353
D_x	167	165	243	264	240	260	298	311
±	21	8		17		7		
	Cs	Ba	Tl	Pb	Bi	Zn	Cd	Hg
n	13	12.5	12	12	(11)	13	13	15
D_c	178	182	179	162		82	73	34
D_x	175	192	195	≤157	<283	86	69	40
±			4					
	Sc	Ti	V	Cr	Mn	Fe	Co	Ni
n	11	11	11	11	10.5	11	10	10
D_c	197	191	186	197	224	169	268	257
D_x	205	205	209	190	251	148	245	240
±	17	9	7					8
	Pt	Yb	Nb	Mo		Ru	Rh	Pd
n	9.5	14	11	11.5		11	11	9.5
D_c	363	157	257	206		235	226	230
D_x	327	159	>222	211		223	241	234
±		38				15	6	25

in which $\tau^+ = \tau^n - \tau^{n+1} = \tau^{n+2}$, n is the bond-order exponent and s the slope of the bond order–$\Delta d'$ function.

In the calculation of heteronuclear interactions, it is necessary to substitute $R_0 = \sqrt{r_0(1) \cdot r_0(2)}$. A serious complication exists therein that, as for homonuclear interactions, the critical stretch to effect a change of bond order depends on atomic volume and, in addition, also on relative atomic size. The strategy to address this problem was to use known parameters, together with experimental force constants, to calculate effective values of $\Delta d'$ from the expression

$$(\Delta d')^2 = \frac{4.615\tau^+ s}{r_0(1) \cdot r_0(2) \cdot k_r}.$$

The calculated values of $\Delta d'$ for heteronuclear diatomics from period 2

	N	O	F
C	0.72(3)	0.78(3)	0.86(2)
N		0.77(3)	
Be		$0.70(2\frac{1}{2})$	

with bond order in parentheses appear to be well ordered. In the same way, the results for groups 2–3 diatomics are considered sufficiently alike to assume $\Delta d' = 0.047$ as a predictor in calculating unknown k_r for this family, including compounds of N:

	C	N	O
Mg			$0.047(2\frac{1}{2})$
Si			$0.051(2\frac{1}{2})$
P	0.045(2)		0.044(2)
S	$0.046(2\frac{1}{2})$		0.047(2)

A similar regularity which emerges for heteronuclear halides

	Cl	Br	I
F	0.09	0.08	
Cl		0.07	
C	0.08	0.06	0.05

confirms that the algorithm yields reasonable results.

4.2.1 Hydrides

The hydrides of p-block elements represent the best documented set of experimental stretching force constants. The previous strategy yields the surprising result that, as for homonuclear interactions, the effective critical stretch $\Delta d'$ is a function of only bond order. In the first linear region, $s = 1.3$, the value $\Delta d' = 0.115$, with bond orders as identified in Table 11, predicts the following force constants:

BH	$k_r = 3.28$	$k_x = 3.05\,\text{N cm}^{-1}$
CH	4.39	4.48
NH	4.62	5.97
OH	7.92	7.80
FH	9.64	9.66

With the exception of NH, this is considered excellent agreement.

In the second period, $\Delta d' = 0.062$ predicts k_r in exact agreement with experiment: SH= 4.22, PH= 3.21 N cm^{-1}. Calculating back from the observed k_r for the hydrogen halides, F to I, the sequence $\Delta d' = 0.115, 0.073, 0.054, 0.046$ appears reasonable.

These results are in line with the small ionization radius of hydrogen, which shows that its entire charge sphere becomes embedded into a larger sphere on molecular formation. The effective point position of the proton relative to the wave structure of the larger atom decides the bond order.

4.3 Molecular Hydrogen

For H alone, the ionization radius is known as an analytical result, $r_0 = 1.835 \times 0.53 = 0.97$ Å. Experimental parameters for H_2 are

$$D_x = 436\,\text{kJ mol}^{-1}$$
$$d = 0.74\,\text{Å}$$
$$k_r = 5.75\,\text{N cm}^{-1}$$

The interatomic distance defines $d' = 0.74/0.97 = 0.763$, which corresponds precisely to bond order 2, with the interesting corollary $d' = 2\tau^2$. Sadly, as inferred from simple number theory, the $D_c = 1,389 r_0^2 \tau^2 = 499\,\text{kJ mol}^{-1}$ does not match the observed.

The factor $\tau^{2.28} \simeq \frac{1}{3}$ produces the observed D_x. It is noted that another molecular form, the familiar H_2^+, has $D_x = 269 \simeq 1,389 r_0^2/5 = 261$ kJ, and with $d' = 1.09$, bond order $b < 0$. The intermediate first-order excited state with $d = 0.84$ Å and $D = 1,389 r_0^2/4 = 327\,\text{kJ mol}^{-1}$ has not been observed.

The observed deviation of the golden exponential from the expected integral value of 2 suggests some exclusion principle that prevents exact second-order interaction. It is shown in Fig. 6 how the superposition of two spherical waves of wavelength $\lambda = a_0/2$ predicts constructive interference at $3\lambda = 0.795$ rather than 0.74 Å, required for $b = 2$. The mismatch is responsible for the shift to $\tau^{2.28} \simeq \frac{1}{3}$, which in real terms is an almost imperceptible deviation from second-order interaction.

An obvious simulation of the stretching force constant for H_2 would be in terms of the stretch $\Delta d' = 1.09 - 0.763 = 0.327$, $\tau^{4.28} \simeq \tau^2/3 = 0.1275$ and unit initial slope:

$$k_r = \frac{4.615 \times 0.1275}{(0.97 \times 0.327)^2} = 5.85\,,$$

compared to $k_x = 5.75\,\text{N cm}^{-1}$.

In order to understand the results reported here, it is imperative to abandon the idea of bond order as a function of electron pairs. The alternative definition in terms

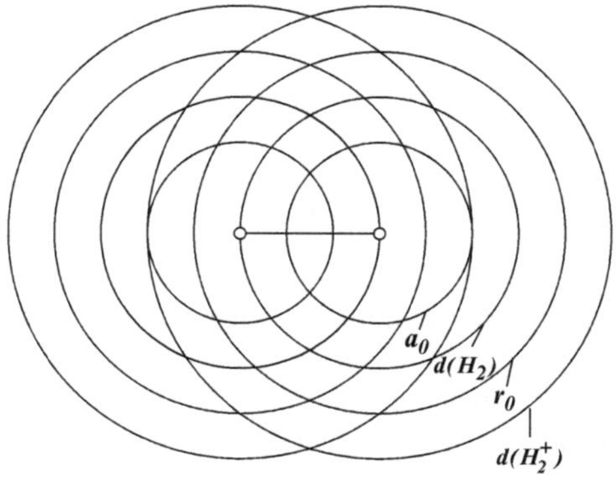

Fig. 6 Graphical simulation of the *H–H* interaction

Table 13 Variation of parameters with bond order

		H	B	C	N	O	F
H	D	436	340	338	<339	430	570
	d	0.74	1.23	1.12	1.04	0.97	0.92
	k_r	5.75	3.05	4.48	5.97	7.80	9.66
	b	2	$\frac{1}{2}$	1	$1\frac{1}{2}$	$1\frac{1}{2}$	2
	n	2.28	7	6.5	6.5	5.5	4.5
F	D	570	732	514	<349	220	159
	d	0.92	1.26	1.40	1.41	1.42	1.41
	k_r	9.66		7.42			4.70
	b	2	2	1	1	$\frac{1}{2}$	$\frac{1}{2}$
	n	4.5	4.5	5	5.5	6	6

of wave structures clarifies many a puzzling feature of conventional theory, such as the relative strengths of electron-pair bonds of the same order. By inspection, the variation of experimental covalence parameters D, d and k_r, shown in Table 13, are qualitatively at variance with a constant bond order of unity as required by the electron-pair model.

However, in terms of the alternative bond order parameters b and the golden exponents n, these quantities are correctly simulated at the quantitative level. The mystery that surrounds many other observations such as the paramagnetism of molecular oxygen or the non-existence of a P_2 triple bond also disappears.

5 Diatomic Dipole Moments

It has been shown that the electronic charge distribution in an atom is readily calculated by the same optimization procedure, based on a golden spiral [6], that correctly predicts all satellite orbits in the solar system [27]. The simulation is

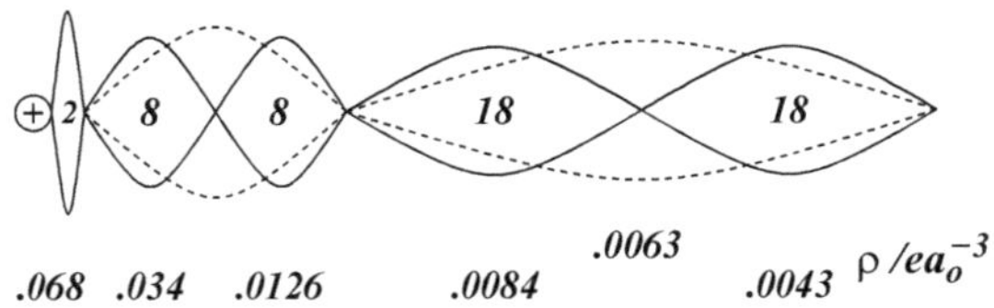

Fig. 7 Electron charge densities in spherical-wave model of the atom

sufficiently reliable to enable an improved derivation of the ionization radii of compressed atoms [18], known to generate a self-consistent electronegativity scale [19]. Having demonstrated that the same elements of number theory also dictate the details of atomic periodicity and the nature of covalent interaction, without the use of higher mathematics, it is of interest to also explore the feasibility of calculating molecular dipole moments by the same approach. The results for diatomic molecules are compared to the data tabulated in HCP [7].

5.1 *The Algorithm*

The formation of a diatomic molecule involves the interaction between two activated valence electrons. In the case of heteropolar interaction, the difference in quantum potential energy (electronegativity) of these two electrons results in a skewed charge distribution, which may be expressed as a difference δQ in charge, measured at the nuclear sites.

The dipole moment of a diatomic molecule is defined as the product of equal, but electrically opposite, fractional charges at the interatomic distance d apart, i.e.

$$\mu = \delta Q \cdot d \; .$$

The magnitude of δQ depends on the differences in atomic electronegativities, polarizabilities and valence densities. The effect of different electronegativities has been considered in the calculation of dissociation energy and is introduced here for a single electron as

$$p^+ = \sqrt{r_0^3(1)/r_0(2)} \qquad \text{together with} \qquad p^- = \sqrt{r_0^3(2)/r_0(1)} \, ,$$

the reverse polarization. This defines the polarization factor $\alpha = p^+/v_1 - p^-/v_2$, where v_i is the number of valence electrons on atom i.

In order to estimate atomic polarizabilities, it is noted that the inverse of charge density at the crests of the spherical-wave representation of atoms, in units of a_0^3/e, should be such a measure. This quantity has been calculated before [6] from a spherical standing-wave model of the atom, shown schematically as a radial projection in Fig. 7.

Table 14 Numerical relationship between ionization radii and atomic polarizabilities

									Ratio
							H		$f(1:8)$
$\frac{15}{r_0}$							15.5		
	Li	Be	B	C	N	O	F	Ne	
$\frac{30}{2r_0}$	6.4	7.0	8.0	8.4	8.9	9.4	9.9	10.4	
									1.63
	Na	Mg	Al	Si	P	S	Cl	Ar	
$\frac{80}{3r_0}$	8.6	9.6	9.7	10.2	10.6	10.8	11.6	12.1	
									1.40
	K	Ca	Ga	Ge	As	Se	Br	Kr	
$\frac{120}{4r_0}$	8.6	9.1	10.0	10.4	10.7	11.1	11.5	11.9	
									1.38
	Rb	Sr	In	Sn	Sb	Te	I	Xe	
$\frac{160}{5r_0}$	8.4	9.0	9.7	10.0	10.4	10.7	11.1	11.6	
									1.38
	Cs	Ba	Tl	Pb	Bi	Po	At	Rn	
$\frac{240}{6r_0}$	9.9	10.5	11.7	12.0	12.4	12.8	13.2	13.7	
									1.38

Ionization radii have been calculated by the spherical-wave model [6]

The reciprocals come out as

$$\frac{1}{\rho_n} = 14.7,\ 29.42,\ 79.37,\ 119.04,\ 158.73,\ 232.55$$

These numbers are converted into a series of integers

$$\frac{1.02}{\rho_n} \simeq 15,\ 30,\ 80,\ 120,\ 160,\ 240\ = m$$

which will be used as coefficients to calculate polarizabilities directly from ionization radii, noting that the quantity $k = m/nr_0$ oscillates about $k \simeq 10$ for period number n, as shown in Table 14.

In this table, units of a_0^3/e and Å^{-1} are mixed. To ensure consistency, we convert 1Å^3 $= (0.52)^{-3}a_0^3 = 7.11a_0^3$. Also, calculated charge densities refer to fully occupied energy levels, whereas the $1/r_0$ simulation specifies one-electron densities. The ratio $f(1:8) \to 1.38$ compensates for this effect to give the complete conversion factor $k = 7.11 \times 1.38 = 9.81$ that generates the numbers $a_n = m/10$ from $1/\rho_n$. From these numbers, polarizabilities are calculated as $s_n = a_n/r_0(n) = f/10$ of Table 14. This result provides a simple conversion of calculated dipole moments into Debye units. Noting that

$$1\,D \equiv 3.336 \times 10^{-30}\,\mathrm{C\,m}$$
$$1e\,\text{Å} = 4.8\,\mathrm{D},$$

the dimensional constant for conversion of dipole moments into Debye units follows directly as $K = 10/3.336 = 3$.

5.2 Results

5.2.1 Interactions of *s–p* Type

Valence density depends on the periodic position of an atom, shown for representative elements in Table 14. The simplest situation to model is the polarization that occurs in an alkali halide molecule, also responsible for the largest dipole moments of diatomic molecules. In effect, a singly charged valence shell interacts with a single vacancy in the valence shell of the halogen atom. The polarization of the alkali shell should decrease with atomic size, which is measured by the period number of the valence shell. The implied decrease in valence density from Li to Na, of $8.6/6.4 \sim 3/2$, suggests $\nu = 1/n$ as approximate scale factor, which could be complicated by the appearance of d and f sublevels. It is a complementary vacancy density that should be taken into account.

The feasibility of these assumptions is validated by calculating a dipole moment for LiF with $n = 2$, $s = 3/\big(2r_0(F)\big) \simeq 1$, $K = 3, \alpha = 2.77, d = 1.56\,\text{Å}$:

$$\mu_c = 3 \times \frac{1}{2} \times 2.77 \times 1.56 = 6.48\,,$$

compared to the experimentally measured $\mu_x = 6.33\,\mathrm{D}$. The assumptions also predict the dipole moments of other alkali fluorides with remarkable accuracy. Using $\delta q = 3/n$ for Na to Cs calculates

$$\mu(\mathrm{NaF}) = (\nu = 1)(n = 3) \times (\alpha = 4.26) \times (d = 1.93) = 8.2 \quad (\mu_x = 8.2\mathrm{D})$$
$$\mu(\mathrm{KF}) = 0.75 \times 5.17 \times 2.17 = 8.4 \quad (\mu_x = 8.6\mathrm{D})$$
$$\mu(\mathrm{RbF}) = 0.6 \times 5.89 \times 2.27 = 8.0 \quad (\mu_x = 8.5\mathrm{D})$$
$$\mu(\mathrm{CsF}) = 0.5 \times 6.43 \times 2.35 = 7.6 \quad (\mu_x = 7.9\mathrm{D})$$

For the higher halides of Li and Na, it is adequate to assume $\delta q = 3.5/n$, in line with Table 14.

A better simulation for the higher fluorides ($n > 3$) is obtained by the physically more sensible assumption of $\delta q \propto V_0(\mathrm{M})/V_0(\mathrm{F})$, e.g. $s(\mathrm{KF}) \propto [r_0(\mathrm{K})/r_0(\mathrm{F})]^3 = 12.2$. Noting the factor n^2 that defines wave nodes, we calculate $\delta q = s/n^2$, i.e.

Table 15 Calculated and observed dipole moments (Debye) of the alkali halides

	F	Cl	Br	I
Li (μ_c)	6.5	7.2	7.0	7.6
μ_x	6.3	7.1	7.3	7.4
Na	8.2	9.1	8.8	8.9
	8.2	9.0	9.1	9.2
K	8.5	10.8	10.9	11.2
	8.6	10.3	10.6	$\sim$ 10.8
Rb	8.4	10.7	10.8	11.2
	8.5	10.5		$\sim$ 11.5
Cs	7.9	10.1	10.2	10.6
	7.9	10.4		

$$\delta q(\text{KF}) = 0.76, \quad \mu_c = 8.53\text{D}$$
$$\delta q(\text{RbF}) = 0.63, \quad \mu_c = 8.42\text{D}$$
$$\delta q(\text{CsF}) = 0.52, \quad \mu_c = 7.86\text{D}$$

Deviations from the ideal rule

$$\frac{3}{2r_0(F)} = \frac{8}{3r_0(Cl)} = \frac{12}{4r_0(Br)} = \frac{16}{5r_0(I)} = 1$$

i.e.

$$\frac{3r_0(Cl)}{r_0(F)} = \frac{16}{3}, \quad \frac{4r_0(Br)}{r_0(F)} = \frac{24}{3}, \quad \frac{5r_0(I)}{r_0(F)} = \frac{32}{3}$$

define the factors that convert volume ratios to the scale, fixed before by

$$\frac{3}{2r_0(F)} \quad \text{as} \quad \frac{nr_0(X)}{r_0(F)}, \quad \text{e.g.} \quad \frac{3r_0(Cl)}{r_0(F)} = 4.54,$$

with factors for Br = 6.87 and I = 9.47. This way we find

$$\delta q(\text{LiCl}) = 1.08 \times 4.54/4 = 1.23, \quad \mu_c = 5.1 \quad (\mu_x = 7.1\text{D})$$
$$\delta q(\text{NaCl}) = 1.22, \quad \mu_c = 9.5 \quad (\mu_x = 9.0\text{D})$$
$$\delta q(\text{KCl}) = 1.00, \quad \mu_c = 10.8 \quad (\mu_x = 10.3\text{D})$$
$$\delta q(\text{RbCl}) = 0.83, \quad \mu_c = 10.7 \quad (\mu_x = 10.5\text{D})$$
$$\delta q(\text{CsCl}) = 0.68, \quad \mu_c = 10.1 \quad (\mu_x = 10.4\text{D})$$

As before simple scaling works better for Li and Na. Final results for all alkali halides are collated in Table 15. The next group of diatomic molecules with

Table 16 Calculation of dipole moments of group 3 fluorides

	$s = \frac{1}{m}V(\mathrm{M})/V(\mathrm{F})$	δQ	α	d (Å)	μ_c	μ_x(D)
BF	1.89/8	0.24	1.72	1.26	0.5	0.5
AlF	5.89/22	0.27	3.45	1.65	1.54	1.53
GaF	7.69/22	0.35	3.99	1.77	2.47	2.45
InF	10.33/28	0.37	4.67	1.99	3.44	3.40
TlF	11.49/28	0.41	4.95	2.08	4.22	4.23

non-trivial dipole moments is the alkaline-earth chalconides where polarization involves double the number of valence electrons and vacancies. Taking this into account, we calculate dipole moments in reasonable agreement with the experimentally known values for MgO = 6.1, SrO = 8.9, BaO = 8.0 and BaS = 10.9 D. Defining $\alpha = p^+/2 - p^-/6, \delta q = (1/n^2)[V(M)/V(O)] \times 4$ we find

$$\mathrm{MgO}: \quad \mu = 2.32 \times (\alpha = 1.62) \times (d = 1.75) = 6.2\mathrm{D}$$

$$\mathrm{SrO}: \quad \mu = 1.76 \times 2.47 \times 1.92 = 9.0\mathrm{D}$$

$$\mathrm{BaO}: \quad \mu = 1.48 \times 2.76 \times 1.94 = 8.0\mathrm{D}$$

For BaS with $3r_0(S)/r_0(O) = 4.63$, $s = (1/36)[V(\mathrm{Ba})/V(\mathrm{S})] \times 4$:

$$\mathrm{BaS}: \quad \mu = 1.87 \times 2.02 \times 2.51 = 9.5\mathrm{D}$$

The main objective is not to produce exact dipoles moments, using fine-tuned parameters, but rather to demonstrate that a convincing match with experimental measurement can be achieved by the multiplication of three factors that derive from ionization radii and valence densities alone.

5.2.2 The *p*-Block Diatomics

In the case of group 3 halides, only p electrons are involved in the interaction. The halogen vacancy is 1/5 of the p-density. Hence, we calculate the polarizability factor as $\alpha = p^+ - p^-/5$. Whereas the s-density of groups 1 and 2 interacts directly with vacancies in the valence shell, the p-density of groups 3 and 7 atoms are not separated by a closed-shell arrangement, but by intervening d-levels. The number of charges that separates B from Al = 8, Al–Ga = 18, Ga–In = 18 and In–Tl = 32. Between groups 3 and 7, there are always ± 4 charges. Instead of scaling the volume ratios by $1/n^2$, the effective scale factors are $m = 8$, $18 + 4$, $32 - 4$, as in Table 16. In the same calculation for some chlorides with known dipole moments, a better fit is obtained with $m \simeq 31$, i.e.

Table 17 Dipole moments of group 4 oxides

	s	δq	α	d (Å)	μ_c	μ_x (D)
CO	1.38/9	0.153	0.72	1.13	0.12	0.11
SiO	4.39/9	0.488	4.20	1.51	3.09	3.10
GeO	5.89/16	0.368	5.38	1.62	3.21	3.28
SnO	7.93/23	0.345	6.74	1.83	4.25	4.32
PbO	8.93/28	0.319	7.34	1.92	4.50	4.64

$$\mu(\text{InCl}) = 2.98 \times 4.54/31 \times (\alpha = 3.59) \times (d = 2.4) = 3.76\,, \quad (\mu_x = 3.79\text{D})$$

$$\mu(\text{TlCl}) = 0.49 \times 3.81 \times 2.48 = 4.59 \quad (4.54)$$

$$\mu(\text{TlI}) = 1.69 \times 9.47/31 \times 3.21 \times 2.81 = 4.66 \quad (4.61)$$

However, the sample is too small to reveal a logical pattern.

Simulation of the dipole moment of CO needs special care. In order to simulate the $2\frac{1}{2}$-order interaction, it is necessary to modify the characteristic radii to r_0(C)=1.784 Å, r_0(O)=1.36 Å, with the number of valence electrons $\nu_C \simeq \nu_O \simeq 3$, to give a polarization factor of $\alpha \leq 0.86$, $\delta q = (1.784/1.36)^3/16 = 0.14$, $d = 1.13$ Å, $\mu_c \leq 0.14$ D ($\mu_x = -0.11$ D). The charge flow from O→C, implied by the modified radii, inverts the sign of the dipole moment as observed.

In comparison, ab initio SCF calculation of μ(CO) with the correct sign, at −0.077 D, requires a double-zeta-plus-polarization basis set with 138 doubly excited configurations plus 62 single excitations [30]. The chemical principles involved here are hard to visualize.

Several chalconide diatomics of the carbon group are also subject to similar modification of their atomic valence spheres. Since these effects have not been calculated, an approximation, which assumes a polarization factor of $\alpha = 2(p^+ - p^-)$, calculated with unmodified atomic radii, with the special scale factors shown in Table 17, was found to give results in good agreement with experiment.

In modeling the higher chalconides, a factor $n r_0(\text{X})/r_0(\text{O})$ scales the fractional charges $\delta q = s \cdot f(\text{X})$, where f(S)= 4.63, f(Se)= 6.78 and f(Te)= 9.32, as in Table 18. For selenides and tellurides, the empirical fit to known dipole moments are

$$\text{CSe}: \quad \mu_c = (0.28 \times 6.78/6) \times (\alpha = 3.80) \times (d = 1.68) = 2.02 \quad (\mu_x = 1.99\text{D})$$

$$\text{GeSe}: \quad \mu_c = (1.6 \times 6.78/10) \times 0.72 \times 2.13 = 1.66 \quad (\mu_x = 1.65)$$

$$\text{GeTe}: \quad \mu_c = (0.97 \times 9.31/12) \times 0.36 \times 2.34 = 1.06 \quad (\mu_x = 1.06)$$

The m index that correlates separated fractional charges with volume ratios is an integer that changes in a regular, but still unspecified way, as in

$$\delta q = \frac{1}{m}[V(M)/V(X)]\,,$$

Table 18 Dipole moments of group 4 sulphides

	s	δq	α	d (Å)	μ_c	μ_x (D)
CS	0.37/4	0.43	2.80	1.54	1.85	1.96
SiS	1.19/4	1.38	0.60	1.93	1.60	1.73
GeS	1.60/13	0.57	1.70	2.01	1.95	2.00
SnS	2.15/20	0.50	2.92	2.21	3.23	3.18
PbS	2.43/25	0.45	3.44	2.29	3.55	3.59

with the relative periodic positions of M and X. The pattern is summarized in the following array:

			F	F	O	Cl	Cl	S	Br	Br	Se
Li	B	C	4	4	9	4		4	4		6
Na	Al	Si	9	22	9	9		9	9		
K	Ga	Ge	16	22	16	16		13	16		10(12Te)
Rb	In	Sn	25	28	23	25	31	20	25		
Cs	Tl	Pb	36	28	28	36	31	25	36		

For alkali halides and alkaline-earth chalconides, $m = n^2$, where n is the period number of M. Where both atoms are in the p-block, m also depends on the periodic position of X.

The dipole moments of the six interhalogen diatomics are modeled well by calculating $\mu_c = \alpha \cdot \delta q \cdot d$, $\alpha = p_+ - p^-$, $\delta q = (3m/20)(V_1/V_2) \equiv 0.15 \times V_1/V_2 \times m$. Hence,

$$\mu(\text{FCl}) = 1.59 \times (0.15 \times 0.289 \times 8) \times 1.63 = 0.90 \quad (\mu_x = 0.89\text{D})$$
$$\mu(\text{FBr}) = 2.26 \times (0.15 \times 0.198 \times 12) \times 1.76 = 1.42 \quad (1.42)$$
$$\mu(\text{FI}) = 2.86 \times (0.15 \times 0.147 \times 16) \times 1.91 = 1.93 \quad (1.95)$$
$$\mu(\text{ClBr}) = 0.62 \times (0.15 \times 0.684 \times 4) \times 2.14 = 0.54 \quad (0.52)$$
$$\mu(\text{ClI}) = 1.16 \times (0.15 \times 0.509 \times 6) \times 2.32 = 1.23 \quad (1.24)$$
$$\mu(\text{BrI}) = 0.55 \times (0.15 \times 0.744 \times 5) \times 2.47 = 0.76 \quad (0.73)$$

Diatomic oxygen halides are correctly modeled by assuming $\alpha = p^+ - p^-$, $\delta q = (V_1/V_2) \times 0.18m$, $m(\text{Cl}) = 10$, $m(\text{Br}) = 12$, $m(\text{I}) = 16$, i.e.

$$\mu(\text{ClO}) = 1.43 \times 0.607 \times 1.57 = 1.36 \quad (\mu_x = 1.30\text{D})$$
$$\mu(\text{BrO}) = 2.08 \times 0.497 \times 1.72 = 1.78 \quad (1.76)$$
$$\mu\text{IO} = 2.67 \times 0.492 \times 1.87 = 2.46 \quad (2.45)$$

The scale factors that convert halogen interactions to the F-scale suggest an index $m < 2$ for OF. A measured value very close to zero is reported. From this, we infer

Table 19 Calculation of hydride dipole moments

	C	N	O	F
v	2	3	4	5
α	2.41 − 0.72 1.69	2.23 − 0.73 1.50	2.05 − 0.76 1.29	1.90 − 0.77 1.12
$u = (V_H/V_X)$	0.16	0.19	0.22	0.26
$s = u/\sqrt{v}$	0.11	0.11	0.11	0.116
$\delta q = ms$	0.77	0.88	1.32	1.74
m	7	8	12	15
d (Å)	1.12	1.04	0.97	0.92
$\mu = \alpha \cdot \delta q \cdot d$	1.46	1.37	1.65	1.78
μ_x (D)	1.46	1.39	1.66	1.83

$m \simeq 0.1$. The same m should model the dipole moment of SF. We find

$$\mu(\text{SF}) = 1.73 \times (12 \times 0.23 \times 0.1) \times 1.60 = 0.76 \quad (\mu_x = 0.79\text{D})$$

Other interactions in the p-block are empirically modeled by a closely related scheme. We find

$$\mu = \alpha \times \delta q \times d$$

$$\mu(\text{NO}) = 0.18 \times (4 \times 0.849 \times 0.18) \times 1.15 = 0.13 \quad (\mu_x = 0.16\text{D})$$

$$\mu(\text{NS}) = 1.59 \times (12 \times 0.320 \times 0.20) \times 1.49 = 1.82 \quad (1.81)$$

$$\mu(\text{OS}) = 1.78 \times (12 \times 0.272 \times 0.18) \times 1.48 = 1.55 \quad (1.55)$$

$$\mu(\text{OP}) = 1.86 \times (12 \times 0.259 \times 0.22) \times 1.48 = 1.88 \quad (1.88)$$

$$\mu(\text{NP}) = 1.67 \times (19 \times 0.305 \times 0.19) \times 1.49 = 2.74 \quad (2.75)$$

5.2.3 Diatomic Hydrides

To find a formula for the dipole moments of diatomic hydrides, we look at the hydrides of the first short period and calculate

$$\alpha = \frac{r_0^3(\text{X})}{r_0(\text{H})} - \frac{r_0^3(\text{H})}{r_0(\text{X})} \quad \text{and} \quad s = V(\text{H})/V(\text{X})$$

as shown in Table 19. The values of m are clearly derived from $v \times 3$, which is exact for HF [$3/2r_0$(F)$\simeq$ 1] and decreases slightly to the left, noting that r_0(C)/r_0(F)$\simeq$ $7/3v$.

By the same method, scaling by the factors 4.54 and 4.63, established before and noting the difference of 2 in the number of vacancies, we calculate

$$\mu(\text{HCl}) = (\alpha = 2.91) \times \left(\frac{1}{\sqrt{5}} \times 0.341 \times 2\right) \times (d = 1.27) = 1.11 \quad (\mu_x = 1.11\text{D})$$

$$\mu(\text{HS}) = 3.33 \times \left(\frac{1}{2} \times 0.278 \times 4\right) \times 0.97 = 1.80 \quad (\mu_x = 1.83\text{D})$$

In the same way,

$$\mu(\text{HBr}) = 3.69 \times (1/\sqrt{5} \times 0.51 \times 6.87) \times 1.41 = 0.82 \quad (\mu_x = 0.83\text{ D})$$

$$\mu(\text{HI}) = 4.40 \times (1/\sqrt{5} \times 0.038 \times 9.47 \times 2/5) \times 1.61 = 0.46 \quad (\mu_x = 0.45\text{ D})$$

The effective scale factors for hydrogen halides are

	HF	HCl	HBr	HI
	15	9.08	6.87	3.80
3×	5	3.03	2.29	1.27

On comparison with similar factors for period 2 hydrides, periodic scaling with respect to F is seen to be such that the cross products between these factors as they appear, moving towards C and I, respectively, are simple multiples of 3, as in the following array:

	C(2.33)	O(2.67)	N(4.0)	
~	3	6	12	15 F
	I(1.27)	Br(2.29)	Cl(3.03)	

This regularity is the result of a periodic relationship between atomic ionization spheres, also manifested in atomic electronegativities.

5.3 Discussion

The calculation of dipole moments described here differs from all other methods in ignoring nuclear charge. The rationale behind this is that any atom is electrically neutral. During covalent interaction, only the extranuclear charge clouds are subject to polarization, which renders heteronuclear diatomics dipolar. As the characteristics of atomic charge clouds are fully characterized by ionization radii and the number of valence electrons, these are the only parameters needed for the calculation of dipole moments of atomic pairs of known periodic positions. Some of the empirical factors introduced here, although poorly understood, are consistent with a regular periodic pattern.

It is only in the case of the alkali halides that a regular pattern in the variation of dipole moment can be identified and interpreted, with some imagination, in terms of the periodic variation of ionization radii. The paucity of data for other heteropolar combinations prevents generalization of the observed trend. There are some tantalizing indications that implicate the role of intervening transition and inner-transition levels, but to a large extent, each dipole calculation still represents a special case. Realizing that for only about 5% of the possible heteronuclear combinations between representative elements have dipole moments been measured, the data to substantiate any general simulation are clearly insufficient. However, the limited success demonstrated here confirms that the appropriate parameters for the calculation of dipole moments have been identified, although not necessarily quantified.

6 Conclusion

It would be wrong to interpret this work as an effort to refute the importance of quantum theory for chemistry. It does the opposite, but questions the methodology that developed from a naïve interpretation of three-dimensional wave mechanics to confirm the electron-pair model of Lewis and the molecular structure theory of van't Hoff. Even in terms of the probabilistic interpretation of wave mechanics, a rigid three-dimensionally structured molecule, with its real molecular orbitals, is undefined. A strategy, based on these concepts and which became known as *Quantum Chemistry*, amounts to a disastrous misreading of quantum theory and has no predictive power beyond its classical basis.

To avoid further confusion, it is recommended to use the term *nonclassical theory* instead of the unfortunate *quantum mechanics*. Nonclassical theory became important after the discovery of the electromagnetic field. The summary of Maxwell's field equations in the form

$$\left(\frac{\partial^2}{\partial x^2}+\frac{\partial^2}{\partial y^2}+\frac{\partial^2}{\partial z^2}-\epsilon_0\mu_0\frac{\partial^2}{\partial t^2}\right)\Phi=0$$

resembles a three-dimensional wave equation and was interpreted as such, despite Minkowski's demonstration that it defines a four-dimensional field. Noting that $t/\sqrt{\epsilon_0\mu_0}$ corresponds to a complex space coordinate, $x_0 = it/\sqrt{\epsilon_0\mu_0}$, the field equation becomes

$$\sum_{j=0}^{3}\frac{\partial^2\Phi}{\partial x_j^2}=\square^2\Phi=0 \tag{6}$$

Classical Newtonian mechanics is a subset of this four-dimensional nonclassical field. Solutions of (6) represent what is colloquially known as either special relativity or quantum theory.

As a quantum theory, (6) introduces the angular-momentum-spin function, fundamental to the periodic table of the elements, which has no recognizable basis in three dimensions. Spin is a purely four-dimensional concept without any meaning in the three-dimensional mechanical world of particles. By definition, it defines a mathematically allowed local configuration of four-dimensional space–time. Projected into three-dimensional space, it appears as a wave packet. Like all wave phenomena, it is characterized by discrete variables, observed as quantum numbers. What the philosopher Popper refers to [31] as the 'quantum muddle' arises from assigning quantum numbers to classical three-dimensional mechanical particles. The way in which number theory is used here to simulate chemical behavior is done in the spirit of four-dimensional nonclassical theory. This way an interatomic distance does not represent a "bond length" in a rigid classical molecule, but an equilibrium situation resulting from the constructive interference between nonclassical valence electron waves. The present results do not inaugurate a new chemistry. It is no more than the tip of an iceberg destined to blossom into something meaningful.

Acknowledgements I gratefully acknowledge the influence of Peter Comba, the only internationally renowned scientist who not only indulged but actively encouraged my maverick views for almost two decades.

References

1. Boeyens JCA (1973) J S Afr Chem Inst 26:94
2. Meyer AM (1988) J Mol Struct (Theochem) 179:83
3. Boeyens JCA (1980) S Afr J Chem 33:14
4. Boeyens JCA (2005) New theories for chemistry. Elsevier, Amsterdam
5. Boeyens JCA (2008) Chemistry from first principles. Springer, Dordrecht, Heidelberg, London, New York
6. Boeyens JCA (2013) Atomic structure, Struct Bond 148:71
7. Lide DR (ed) Handbook of chemistry and physics, 86th edn. CRC Press, Boca Raton.
8. Kekulé A (1861) Lehrbuch der Organiaschen Chemie. Enke, Erlangen.
9. Meyer V (1888) Chem Ber 21:964, 1620
10. Stewart AW (1922) Some physico-chemical themes. Longmans, London
11. Sommerfeld A (1924) Atombau und Spektrallinien, 4th edn. Vieweg, Braunschweig
12. Pauling L (1960) Nature of the chemical bond, 3rd edn. Cornell University Press, Ithaca, p 113
13. Boeyens JCA, Schutte CJH (2012) In: Putz MV (ed) Chemical information and computational challenges in 21st century. Nova, New York
14. Putz MV (ed) (2012) Chemical information and computational challenges in 21st century. Nova, New York
15. Boeyens JCA (1982) J Crystallogr Spectr Res 12:245
16. Boeyens JCA, Ledwidge DJ (1983) Inorg Chem 22:3587
17. Boeyens JCA (1978) S Afr J Chem 31:121
18. Boeyens JCA (1994) J Chem Soc Faraday Trans 90:3377
19. Boeyens JCA (2008) Z Naturforsch 63b:199
20. Boeyens JCA, Levendis DC (2008) Number theory and the periodicity of matter. Springer.com
21. Boeyens JCA, O'Neill FMM (1995) Inorg Chem 34:1988
22. Boeyens JCA, O'Neill FMM (1998) Inorg Chem 37:5352
23. Bacsa J, Boeyens JCA (2000) J Organomet Chem 596:159
24. Boeyens JCA (1985) Inorg Chem 24:4149
25. Boeyens JCA, Cotton FA, Han S (1985) Inorg Chem 24:1750

26. Boeyens JCA (2010) Int J Mol Sci 11:4267
27. Boeyens JCA (2009) Phys Essays, 22:493
28. Boeyens JCA (2010) Chemical cosmology. Springer, Dordrecht, Heidelberg, London, New York
29. Cotton FA, Walton RA (1982) Multiple bonds between metal atoms. Wiley, New York
30. Schaeffer HF III (1972) The electronic structure of atoms and molecules. Addison-Wesley, Reading
31. Popper KR (1992) Quantum theory and the schism in physics. Routledge, London

Struct Bond (2013) 148: 137–159
DOI: 10.1007/978-3-642-31977-8_6
© Springer-Verlag Berlin Heidelberg 2013

Molecular Shape

Peter Comba and Jan C.A. Boeyens

Abstract Molecular shape is recognized as an emergent property that complements the projection from four-dimensional space–time to tangent Euclidean space. Projection from hypercomplex algebra to real algebra necessitates the three-dimensional definition of concepts such as chirality, quantum uncertainty and probability density to compensate for errors of abstraction. The emergent alternative description of extranuclear charge density as spherical standing waves, optimized by a golden spiral, reveals atomic structure in line with the periodic table of the elements and underpinning the concepts of bond order, interatomic distance and stretching force constant, related to chemical interaction. The principles giving rise to molecular structure are shown to depend, like bond order, on the constructive interference of atomic wave fields, optimized by minimal adjustment to bond orders. The procedure is shown to be equivalent to the philosophy of molecular mechanics. Arguments based on the traditional interpretation of electronegativity are presented to relate the parameters of strain-free bond lengths, dissociation energies and harmonic force constants, used in molecular mechanics, to quantum mechanically define ionization radii of atoms. Atomic electron densities and a bond-order function, both obtained by number-theory optimization, enable the direct calculation of interatomic distance, dissociation energy and stretching force constant for all pairwise interactions of any order. Torsional interaction determines the final shape of a molecule and presumably can only be understood as a four-dimensional effect.

P. Comba (✉)
Anorganisch-Chemisches Institut, Universität Heidelberg, Im Neuenheimer Feld 270, 69120 Heidelberg, Germany
e-mail: peter.comba@aci.uni-heidelberg.de

J.C.A. Boeyens
Unit for Advanced Scholarship, University of Pretoria, Lynnwood Road, 0002 Pretoria, South Africa
e-mail: jan.boeyens@up.ac.za

Keywords Angular momentum · Emergent property · Molecular mechanics · Molecular structure · Torsional interaction

Contents

1 Introduction

A principal aim of chemical analysis is to develop a theoretical model of the interaction between atoms and molecules. Experimental work of the previous two centuries has resulted in a highly successful empirical account of chemical reactivity, and efforts to formulate a rigorous, fundamental theory as a nonclassical many-body problem have lead to highly accepted and much used methods but these still have significant limitations. By the current approaches, chemical interaction is modeled in terms of probability-density distributions of independent electrons. Although the theory appears to work for one-particle problems, unforeseen effects emerge in the treatment of more complex systems [1]. In particular, the distribution of extranuclear electrons seems to obey an exclusion principle, not anticipated in the basic theory, and there is no fundamental understanding of three-dimensional molecular shapes, as observed experimentally. The pivotal role of entropy, which controls the course of chemical reactions, is theoretically equally unexpected.

It is not unexpected that problems often occur in the fundamental analysis of emergent properties. Maybe the prudent response of the chemist should then be a critical reexamination of those assumptions that underpin the partially successful theory. In any theory, there is a reductionist limit, beyond which there are no data to guide the recognition of more fundamental principles. In the theory of matter, this limit occurs in the vacuum, or sub-ether [2], seen as the primaeval form of matter, continuously spread across the endless void. On deformation of this featureless cosmos, ponderable matter emerges from the void as elementary distortions, which are perpetually dispersed, except in a closed system. We propose such a structure as the primary assumption in the theory of molecular shape and assume that persistent elementary units occur in the form of what will be called waves, for lack of better terminology. The elementary waves exhibit the first emergent properties of mass, charge and spin, which they possess in characteristic measure.

Matter in all forms can now be recognized as consisting of the three robust elementary forms known as proton, electron and neutrino. A fourth common form, known as neutron, only occurs in close combination with protons. It has a limited lifetime in free space where it decays into an electron, a proton and a neutrino.

The postulated elementary units should not be confused with waves as perceived in three-dimensional space, but rather as undulations in four-dimensional space–time. The mathematical description of these two types of swaying is fundamentally different. A steady state that results from four-dimensionally balanced forces, as a function of a scalar potential Φ, obeys an equation of the type

$$\frac{\partial^2\Phi}{\partial x_0^2}+\frac{\partial^2\Phi}{\partial x_1^2}+\frac{\partial^2\Phi}{\partial x_2^2}+\frac{\partial^2\Phi}{\partial x_3^2}=0 \tag{1}$$

By assuming Φ in the form of a product function, each term in

$$\frac{1}{\Phi}\Box^2\Phi=\sum_{\mu=0}^{3}\frac{1}{X_\mu}\frac{\partial^2\Phi}{\partial x_\mu}=\sum k_\mu^2=0$$

defines a characteristic constant on condition that $\sum k_\mu^2=0$. The general solution of (1) is thereby defined as a four-dimensional hypercomplex function, known as a quaternion, in which all variables remain inextricably entangled.

A common approximation that reduces the equation into a three-dimensional wave equation assumes the separation of space and time coordinates, which is the basis of wave mechanics. For many purposes, this is a good approximation in tangent Euclidean space, but it has no validity in curved four-dimensional space–time.

It is important to note that the property of spin is only defined in quanternion notation, which specifies a conserved quantity $\mathbf{J}$. It may be viewed as a four-dimensional symmetry operator, approximated by a three-dimensional angular-momentum operator $\mathbf{L}$ and a one-dimensional spin, on separation of space and time variables. The approximation $\mathbf{J}=\mathbf{L}+\mathbf{S}$ implies that neither $\mathbf{L}$ nor $\mathbf{S}$ is a three-dimensional vector, both of them implying rotation in spherical mode [3]. The one-dimensional projections, $\mathbf{L}_z$ and $\mathbf{S}_z$, in an applied magnetic field or in a molecular environment are vector quantities.

2 Space-Like Correlations

It is appropriate to digress at this point into a discussion of the much-debated non-locality of quantum theory. In a nutshell, it amounts to the observation that a wave-mechanical correlation, established at a given point, remains operational even as a correlated pair drifts apart indefinitely. The logical implication of this is instantaneous action at a distance, an idea much maligned over the ages but well defined as space-like interaction in the theory of special relativity. The interval between four-dimensional points in Minkowski space,

$$ds = \sqrt{dx_0^2 - dr^2},$$

reduces to zero in the surface of the light cone and becomes complex on the outside. Whereas a stationary object within the light cone moves through time with constant space coordinates, the time coordinate remains constant for such an object in the space region. Irrespective of spatial separation, any pair of points in four-dimensional space–time therefore remains in virtual contact and correlated.

Humans are conditioned to interpret the environment in three- rather than four-dimensional detail and destined to experience events that appear normal in four dimensions as mysterious in three. Common examples include non-locality, the commutation properties and other mysterious features of wave-mechanical variables and the chirality of space.

3 Wave-Mechanical Approximation

The nonclassical mathematical description of the world follows Eq. (1), which in practice is solved by the separation of space and time variables. Although it is a good approximation, it cannot render four-dimensional effects intelligible in three. The problem is highlighted by analogy with efforts to describe geometrical shapes in lower-dimensional space.

Mirror-related triangles in two dimensions clearly define a chiral pair, which appear achiral in three dimensions.

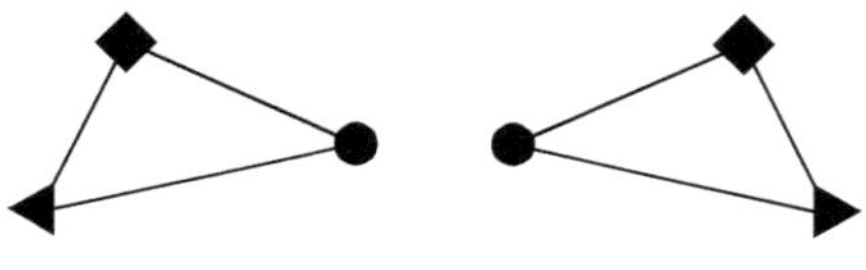

In the same way, the chirality of a three-dimensional tetrahedron is resolved in four dimensions, which means that the three-dimensional chiral forms are identical when described four dimensionally. Small wonder that all efforts to find a wave-mechanical difference between *laevo* and *dextro* enantiomers are inconclusive. The linear superposition principle, widely acclaimed as a distinctive property of quantum systems, is now recognized as no more than a partially successful device to mimic four-dimensional behavior. This includes one of the pillars of chemical-bonding theory, known as the resonance principle.

Probably the most distinctive feature of quantum systems is the non-zero commutators of conjugate variables, said to represent a drastic departure from classical behavior. In actual fact, this is a standard feature of any $\mathbb{H}^4$ algebra, which only becomes problematic on trying to reformulate this in $\mathbb{R}^3$. All of the foregoing is of decisive importance in a theory of molecular shape and creates a serious interpretational dilemma. The strictly three-dimensional empirical data simply cannot support a quasi four-dimensional theory, whereas many observed features cannot be accounted for classically. The point is that nuclear position, the decisive parameter, is a strictly classical particle property, but the interaction between atomic nuclei is wave-like. The traditional compromise to represent both nuclei and electrons by probability-density functions does not work in a strict sense.

4 Atomic Structure

The formation of molecules is driven by the interaction between the extranuclear electronic charge clouds that surround atomic nuclei. According to the wave model proposed here [4], such an electron cloud is conveniently considered as a spherical standing wave in the form of concentric annular shells. Each shell consists of a fixed number of electrons, proscribed by the quantum numbers l and s. All inner shells in a ground-state atom are considered closed with the maximum number of electrons. Except for some special cases such as the inert gases, the outermost valence shell is, by definition, not a closed shell. However, given the sublevel degeneracy of $2l + 1$, with magnetic quantum numbers in the range $-l \leq m_l \leq l$, the valence shell remains spherical in all cases. The quantum number $m_l = 0$ defines a real wave function with spherical symmetry, and any pair of complex functions corresponding to $\pm m_l$ also defines the same symmetry. The quantum numbers for any number of electrons in the valence shell can therefore always be assigned so as to define spherical symmetry. This stipulation is known as Hund's rule.

It is important to realize that all wave-mechanical predictions pertaining to atomic structure are only valid for the H atom. For this reason, wave mechanics is only partially successful as a descriptor of the periodic table of the elements and, hence, of the electronic configuration of many-electron atoms.

Simulation by number theory is the only known procedure that generates the detailed structure of the periodic table without further assumptions or ad hoc corrections. In its simplest form, the simulation is based on the fact that any atomic nucleus consists of integral numbers of protons (Z) and neutrons (N), such that the ratio Z/N is a rational fraction. This ratio converges from unity to the golden ratio (τ) with increasing atomic number and yields a distribution commensurate with the periodic table. The detailed structure of the periodic function is contained in the Farey sequence $\mathscr{F}_4$ of rational fractions and visualized in its Ford-circle mapping [5].

Noting that the periodic table derives from the extranuclear electronic configuration of atoms, it would seem reasonable to assume that a number-theoretic simulation could reveal this distribution as well. In this instance, we are dealing with the special distribution of matter around an active center, the type of problem amenable to analysis by optimization in terms of logarithmic spirals. The only requirement is recognition of an appropriate convergence angle. Using the maximum valence-shell degeneracy of $2n - 1$ at the principal level n, a distribution that

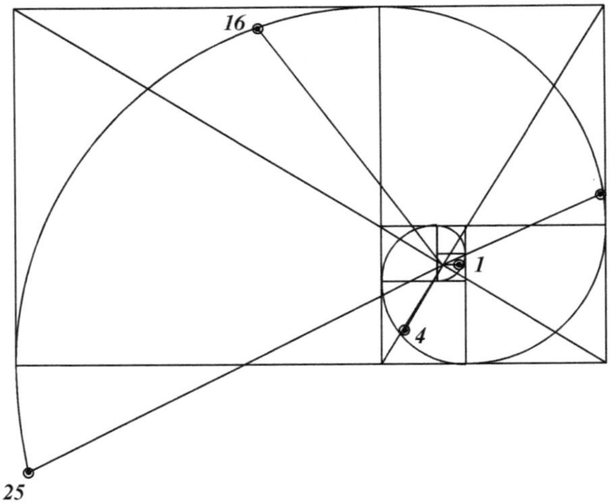

Fig. 1 Points generated in a golden rectangle by a Fibonacci spiral with a variable convergence angle of $4\pi/(2n-1)$. *Numbers* indicate the distance to the spiral center in units of a_0

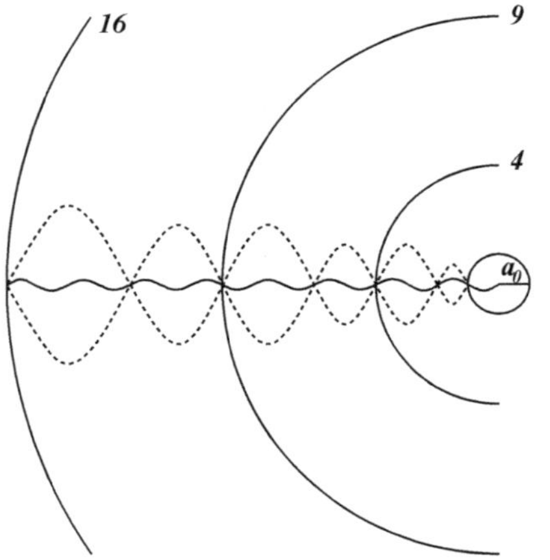

FC	*BS*	*n*
	2 He	1
2 He		
8 (10) Ne		
	8 (10) Ne	2
8 (18) Ar		
18 (36) Kr		
	18 (28) Ni	3
18 (54) Xe		
32 (86) Rn		
	32 (60) Nd	4
32 (118)		
	50 (110)	5

Fig. 2 Atomic shell structure as it emerges from electron-density optimization on a golden spiral. The variable convergence angle of $4\pi/(2n-1)$ manifests in the appearance of $2n-1$ additional cycles (s, p, d, f) in each interval between Bohr levels n and $n-1$, shown here as elementary ripples. In contrast to the Bohr–Schrödinger (BS) model, closed shells in the Ford-circle simulation (FC) invariably coincide with noble-gas configurations

appears to replicate the radii of the Bohr model at n^2a_0 for a variable convergence angle of $4\pi/(2n-1)$ was indicated [6] and is shown here in Fig. 1.

However, in terms of the known periodic structure, the Bohr interpretation cannot apply. In order to generate the periodic table, it is necessary to interpret the extremum condition as satisfied by the nodal surfaces of the spherical electron wave, as shown in Fig. 2. The detailed periodic structure, together with subshell degeneracy of $2n - 1$, corresponding to s, p, d, f spectroscopic states, emerges from the graphical representation. The model has been shown [4] to produce the electronic distribution on all atoms in quantitative detail.

It may be unexpected to find that number theory and traditional wave mechanics yield comparable reconstructions of extranuclear electronic configurations. However, both models are based on classical waves in three-dimensional space, appropriate for the understanding of atomic structure in tangent Euclidean space.

5 Molecular Structure

Towards an understanding of molecular shape, it is important to enquire into the symmetry of an activated atom in a chemically crowded environment. Whether the quantum number m_l represents, as traditionally interpreted, a measure of directed orbital angular momentum, or an element of symmetry, is immaterial. An interatomic collision either redirects the orbital angular momentum of the extranuclear electrons or distorts the symmetry of the charge clouds. The onset of covalent interaction is recognized in either the quenching of orbital angular momentum or, equivalently, optimization of the overall symmetry as specified by Laplace's equation. Either way this principle establishes a criterion in terms of which to predict the relative orientation of sub-molecular fragments that join up in a chemical reaction. Viewing the quantum number m_l as an orbital angular-momentum vector along the polar direction (conventionally denoted by z) is the more convenient practice. It has the advantage that residual angular momentum is the recognized diagnostic of a magnetic moment that generates optical activity.

In the case of second-order covalent interaction, the angular-momentum vectors line up antiparallel in a direction perpendicular to the axis of interaction. The energy which is required to decouple these vectors measures the steric rigidity of the arrangement, known as a barrier to rotation. In terms of the wave picture, steric rigidity relates to the overlap mode of wave crests as shown for ethylene and ethyne in Fig. 3.

The universally accepted model of the second-order dicarbon interaction, colloquially known as an ethylenic double bond, has developed from a misreading of a seminal paper which discussed the quantum theory of double bonds [7]. In this paper, it is shown that by linear combination of the eigenfunctions

$$\psi(r,z)\frac{\mathrm{e}^{\pm i\varphi}}{\sqrt{2\pi}},$$

which describe the angular momentum on a C atom, two eigenfunctions of different energy,

$$\psi(r,z)\frac{\cos\varphi}{\sqrt{\pi}} \quad \text{and} \quad \psi(r,z)\frac{\sin\varphi}{\sqrt{\pi}},$$

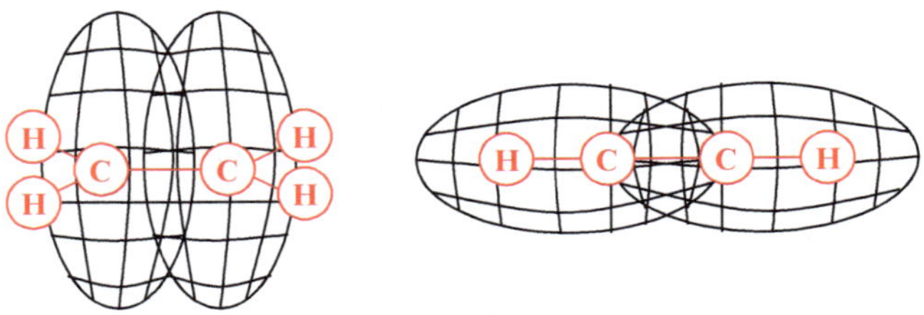

Fig. 3 Relative rotation about the axis of interaction destroys the interference pattern between the wave forms shown on the *left*, but not of those on the *right*

are produced. Interaction between two $-CH_2$ units is next analyzed by perturbation theory in terms of two derived functions of even and odd parity with respect to the plane $\varphi = 0$:

$$\psi_g(r, z, \varphi) \quad \text{and} \quad \psi_u(r, z, \varphi)\,.$$

It was correctly pointed out that with these linear combinations, a moment, as in O_2, no longer exists around the C–C axis. ψ_u^2 is identified as the diatomic ground state which is shown to require a planar C_2H_4 arrangement, interpreted as the cause of the steric rigidity of ethylene.

Derivative work, based on this analysis, erroneously assumes degeneracy of the linear combinations when defining a set of orthogonal real "orbitals". There is no theoretical basis in the seminal paper to justify this assumption. However, what is clearly implied is that the orbital moment on each carbon atom is directed perpendicular to the molecular plane. The molecular angular momentum is therefore quenched vectorially only for this planar arrangement of the molecule. Torsional distortion which creates residual angular momentum therefore requires work, the true basis of a barrier to rotation.

It is instructive to note that the requirement of quenched angular momentum predicts the same tetrahedral geometry for methane [8] as the concept of symmetry optimization.

By exploiting these principles, it becomes feasible to reconstruct the general topology of complicated molecules with known connectivity. Optimization of the topological shape to produce the geometrical details of molecular structure may be done by the methods of molecular mechanics.

During diatomic covalent interaction, the spherical electronic waves on free atoms generate specific interference patterns, which define bond order, interatomic distance and a stretching force constant, characteristic of the interaction. These characteristic properties remain largely intact as the diatomic fragment becomes incorporated in larger molecular assemblies, the three-dimensional structure of which, in the first instance, depends on the interference between second-neighbor waves. It is almost axiomatic that the wave structure of the central atom in a planar arrangement will be elongated as shown for ethylene in Fig. 3.

The geometry of a four-atom molecule, as shown in Fig. 4, but with dissimilar atoms, is specified by three first-neighbor and three second-neighbor interatomic

Fig. 4 Schematic drawing of the interacting spherical electronic waves among four identical atoms in a plane

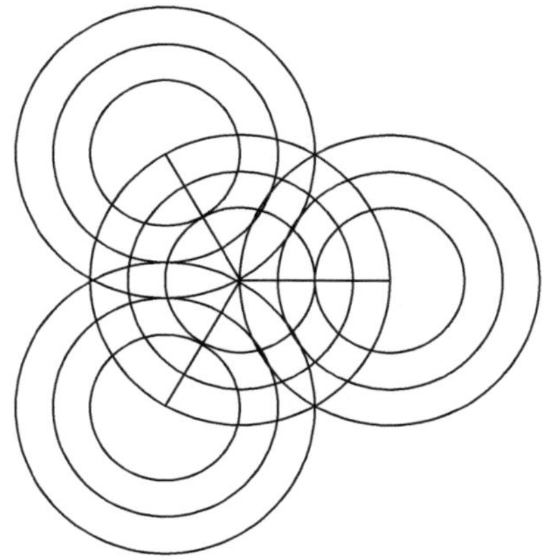

distances—a total of six independent parameters, not necessarily compatible with constructive interference among all wave systems. In order to arrive at an arrangement that optimizes comprehensive constructive interference, it may be necessary to make small adjustments to the six independent bond orders. The required adjustments will be functions of the relative stretching force constants. Each adjustment requires an amount of work,

$$w = \frac{1}{2}k_r\delta^2\,,$$

where δ defines a linear increase in interatomic distance. The optimized nuclear framework will in general not be planar, as suggested by the diagram.

The simple procedure, outlined here and extended over any number of connected atoms, with minimization of the total work required to produce the optimal structure, constitutes the philosophy of *molecular mechanics* (MM). In practice, all secondary interactions are described in terms more familiar to structural chemists. In this way, 1,3-interactions are formulated as deformation of characteristic *valence angles*, according to

$$w_\theta = \frac{1}{2}k_\theta(\Delta\theta)^2\,,$$

based on an *angle-bending* force constant. A 1,4-interaction is reduced to a *torsional* function that describes a periodic barrier to rotation and a so-called *nonbonded* interaction, which is also used to incorporate more remote interactions into the force field. Special parameters are added to deal with electrostatic interactions between polar regions and to maintain the planarity of conjugated systems. Another refinement considers the interdependence of stretches and bends pertaining to common atoms.

Historically, molecular mechanics has developed from a purely empirical procedure to refine molecular trial structures by the minimization of steric energy as a function of nuclear coordinates [9, 10]. The trial structure is generated by assigning empirically idealized interatomic distances and valence angles according to the chemical connectivity pattern. Today, under certain conditions, it is not only the most efficient but also a very reliable procedure for the optimization of three-dimensional molecular structures, not only but specifically also for large systems such as macromolecules, condensed phases and the analysis of conformational space [9, 11].

Although most of the parameters used in MM simulation are based on adequate theoretical concepts, the overall procedure to generate molecular structure remains essentially empirical. The recognized benchmark generally is an experimental structure and therefore not a free molecule but a selected rigid fragment from a bulk structure (i.e. a crystal), and specific intermolecular interactions are generally ignored. It was noted that isotropic (averaged) intermolecular interactions are included in a force field based on crystal structures, and this mimics the situation in solution. Therefore, force fields based on experimental structures are not reproducing structures of isolated molecules and are slightly different from force fields derived from quantum-chemical structures, which model the structures of isolated molecules [9, 11, 12]. As intermolecular interactions are among the weakest forces in bulk phases, bond distances and valence angles may not be significantly affected by the environment. It is mainly the relatively weak torsional interactions that differ significantly from their free-molecule counterparts. Since the torsional flexibility of (primarily carbon–carbon) single bonds is the basis of conformational variations, the shape of a molecule considerably depends on intermolecular interactions. It follows that MM is only able to correctly predict the shape if solvation, hydrogen bonding, long-range van der Waals interactions, electrostatics and polarization are accurately computed and if a full conformational analysis is performed. The same is obviously true for any other structural modeling approach.

5.1 Wave Mechanics

Quantum theory in general has problems similar to those discussed above for molecular mechanics. In addition, the computation of bulk phases and conformational analysis are computationally much more expensive and therefore virtually impossible. Matter, in its most elementary form, is nothing but a special configuration of four-dimensional space–time. The allowed shapes of material aggregates must therefore depend on the topology of space–time, and any theory of molecular matter is inferred to incorporate some aspect of four-dimensional symmetry. However, the wave-mechanical model of matter, on assuming the separation of space and time coordinates, destroys the four-dimensional symmetry, as evidenced by the disappearance of the spin variable that links the shape of material aggregates to the topology of space–time.

In the form of a three-dimensional wave equation, the theory defines a complex variable associated with a vector model of angular momentum. By the principle of symmetry optimization, viz., minimization of angular momentum, this variable may be used to predict the internal three-dimensional symmetry of assembling molecules [8]. Followed by MM optimization, it might constitute a method of predicting classical molecular structure from first principles. However, the standard procedures of *quantum chemistry*, which rely on further separation of space variables, sacrifice the angular-momentum parameter, in order to eliminate the complex variable, and hence suppress the facility to predict internal molecular symmetry.

6 Molecular Mechanics

Among the techniques for structure optimization of chemical systems, molecular mechanics (MM) is by far the fastest and therefore, for large systems such as crystal lattices, polymers, proteins and solutions, often the only useful method, especially when a significant part of the conformational space and/or dynamics need to be included [9, 13]. In many areas, where accurate force fields have been carefully optimized, e.g. for carbohydrates and organic compounds in general [9, 10, 14], but also, e.g. for cobalt(III) hexaamines [9, 15–17], the accuracy of the optimized bond distances is <0.01 Å, and there also is good agreement between computed and experimental thermodynamic properties (relative strain energies) and vibrational frequencies [9, 10, 18, 19], i.e. the MM-derived parameters are nearly as accurate as the experimental data (e.g. X-ray crystallography) and in many cases better than those derived from quantum-chemical methods (QM).

Why then bother about much more expensive QM-based models? One reason is that MM may only lead to accurate results for molecules of the same type used for the optimization and validation of the force field, i.e. extrapolation is seen to be dangerous if not impossible [9]. This also extends to transition states and short-lived, unstable intermediates and therefore to chemical reactivity. Since electrons are not considered explicitly in MM, electronic effects related to structural distortions, specific stabilities and spectroscopy cannot be modeled by MM. However, in all other areas, there is no good reason for not using a well-optimized and validated MM model. Also, there are MM-based approaches to deal with most of the deficiencies listed above [9, 20–28]. In the last decade, there have been a number of approaches, which have, based on simple rules [29], valence bond theory [30–33] and ligand-field theory [20–23], allowed the simplification of the force-field optimization and validation procedures and/or inclusion of electronic effects in MM models.

Therefore, the probably most serious disadvantage of MM compared to all other approaches in structural modeling is a seemingly missing theoretical basis, and the unspoken consensus is that, despite its successes, MM should eventually give way to more sophisticated QM-based models. It is primarily density functional theory (DFT), which in recent years, due to important developments in theory, hard- and software, has taken over some of the ground from MM. An interesting approach related to this observation is that the full potential of DFT is then realized when it is used to build up a force field [34]. Another important aspect, of course, is that MM in general only produces optimized structures and minimized energies, i.e. no information about electronics in ground and excited states.

6.1 Electronegativity

An alternative approach to QM, for the elucidation of the electronic basis of MM, is to reexamine the traditional model of covalent interaction mediated by electron pairs, based on the notion of chemical affinity. The assumption is that the interaction between a pair of atoms in a molecule only involves their valence electrons. If the two atoms are sufficiently alike, equally shared electron density between the atoms binds them together; if they are of different chemical nature, the sharing is unequal and the covalency reduced.

It has been a constant pursuit of chemists to classify the elements in terms of such an affinity factor, in order to predict details of their interaction. The discovery of the periodic table provided the first clue towards identification of such a chemical-bonding parameter, which later became known as electronegativity. On plotting Lothar–Meyer atomic volumes as a function of atomic number, the elements are divided into two kinds, according to the local slope of the connecting curve [35]. This observation led to the classification into electropositive and electronegative elements.

Electronegativity has recently been redefined [36] as the quantum potential of the atomic valence state, calculated from the ground-state energy of an electron, confined to the ionization sphere of radius r_0:

$$E_g = \frac{h^2}{8mr_0^2}$$

To stay in line with common practice, it is convenient to assign a value of $\chi = \sqrt{E_g}$, with E_g in eV, for general use. Numerically this gives $\chi = 6.133/r_0$, with r_0 in Å units.

Since r_0 is characteristic for each atom, characteristic energies are predicted for atomic valence state electrons. It is the atomic equivalent of the Fermi energy of an electron at the surface of the Fermi sea in condensed phases and in that sense represents the chemical potential of the valence electron for each atom. Electronegativity has been defined independently [37] in almost identical terms before. It is a function of only the electronic configuration of atoms and emerges naturally in the response of an atom to its environment. Alternatively, it is the tendency of an atom to interact with electrons and the fundamental property that quantifies chemical affinity and bond polarity.

It is instructive to examine the periodic variation of valence state electronegativities, as a function of atomic number. It separates into the same segments as the Lothar–Meyer curve, and the qualitative trends are recognized as related to the known empirical trends of other electronegativity scales [36]. The slope of the curves at each atomic position represents a change in energy as a function of atomic number (i.e. number of electrons) and defines the chemical potential of the electrons, $\mathrm{d}E/\mathrm{d}n = -\mu$, at that point.

6.2 Simulation by Number Theory

In a previous analysis, based on the generalized covalency curve and empirically adjusted values of ionization radii, r_0 [8], it was shown how to obtain useful MM force-field parameters. An even simpler and more reliable method has now become available by recalculation of atomic ionization radii directly from numerical optimization of valence densities.

The recalculated ionization radii are essentially free-atom values and therefore no longer parametrically related to the general covalence curve. For homonuclear interactions of the same order b, a common dimensionless interatomic distance d_b' is predicted, such that $d = d_b' r_0$. By considering the interaction as an interference between spherical standing waves, integer and half-integer bond-order parameters d_b' are readily optimized with a golden logarithmic spiral. These parameters vary between $d_0' = 1$ and $d_4' = \tau$ to allow the calculation of dissociation energy from a bond order related to some power of the golden ratio [38]:

$$D_c = K r_0^2 \tau^n .$$

K is a dimensional constant. For heteronuclear interaction,

$$D_c = K r_0^3(1)\tau^n / r_0(2); \quad r_0(1) > r_0(2).$$

The optimized quantized values of bond order correspond in an overwhelming majority of cases directly with the traditionally recognized single, double, triple and quadruple bonds that count electron pairs per interaction. Deformation of any diatomic system is resisted by the disturbance of an optimal arrangement and measured with a harmonic stretching force constant. Its value depends on the differences $\Delta D'$ and $\Delta d'$, and the slope, s, of the line that represents the change between bond orders as a continuous function as shown in Fig. 2 of the covalence paper [38]. Calculation of $\Delta D'$ is simplified by a special property of the golden ratio: $\tau^{n+1} - \tau^n = \tau^{n+2} \equiv \tau^+$. In the common units of $\mathrm{N\,cm^{-1}}$ or mdyne/Å, the force constant for a homonuclear stretch is defined as:

$$k_r = \frac{4.615\tau^+ s}{(\Delta d' \cdot r_0)^2}.$$

6.2.1 Nonbonded and 1,3-Interaction

The tetrahedral environment of a covalently saturated carbon atom specifies the separation between 1,3-neighbors by simple trigonometry,[1] with $d(\text{C–H}) = 1.12\,\text{Å}$ and $d(\text{C–C}) = 1.54\,\text{Å}$ as $d(\text{H}\cdots\text{H}) = 1.83\,\text{Å}$, $d(\text{C}\cdots\text{C}) = 2.52\,\text{Å}$, $d(\text{C}\cdots\text{H}) = 2.19\,\text{Å}$.

The H$\cdots$H distance is close to the $2r_0$ nonbonded limit, with an effective bond order of -1, as shown in Fig. 5. For C$\cdots$C, the approximate $d' = 2.52/1.78$ is interpreted to indicate an effective bond order of $-\frac{1}{2}$.

From these data and the slope of 10 in the nonbonded region, 1,3 stretching force constants are calculated and converted into angle-bending constants. In keeping with common MM practice, other nonbonded contacts may be interpreted as van der Waals interactions, but no effort has been made so far to calculate an attractive component. Nonbonded distances of less than the van der Waals limit of $R_{\text{vdW}} = r_0(1) + r_0(2)$ amount to repulsion against a force constant, as shown below. Both 1,3 and van der Waals interactions are simulated by the same function, the latter at somewhat lower order. To calculate k_r for the nonbonded C$\cdots$C interaction, we consider a stretch from bond order $0 \rightarrow -\frac{1}{2}$ ($\Delta d' = 0.5$) for 1,3, and $0 \rightarrow 1$ ($\Delta d' = 1$) for van der Waals interactions. From Table 3 of the Chapter on Covalent Interactions (see p. 102) [38] $n^+ = 10$. For H$\cdots$H and C$\cdots$H, we consider a stretch from $0 \rightarrow 1$ ($\Delta d' = 1$). For 1,3 interactions $n^+ = 9$ and for van der Waals, with $R_{\text{vdW}} > R(1,3)$, we assume $n^+ = 10$. The results under these conditions are summarized by the following calculations:

$$
\begin{aligned}
k_r(\text{C}\cdots\text{C}) &= \frac{4.615 \times \tau^{10}}{(0.5 \times 1.78)^2} \times 10 = 0.47\,\text{N}\,\text{cm}^{-1} \\
k_\theta &= k_r \times 2.52/1.911 \\
&= 0.64\,\text{mdyne/Å rad.} \\
R_{\text{vdW}} &= 3.56\,\text{Å}; \quad k_r = 0.12\,\text{N}\,\text{cm}^{-1}
\end{aligned}
$$

$$
\begin{aligned}
k_r(\text{H}\cdots\text{H}) &= \frac{4.615 \times \tau^{9}}{(1 \times 0.98)^2} \times 10 = 0.63\,\text{N}\,\text{cm}^{-1} \\
k_\theta &= k_r \times 1.83/1.911 \\
&= 0.62\,\text{mdyne/Å rad.} \\
R_{\text{vdW}} &= 1.96\,\text{Å}; (\tau^{10})\, k_r = 0.40\,\text{N}\,\text{cm}^{-1}
\end{aligned}
$$

$$
k_r(\text{C}\cdots\text{H}) = \frac{4.615 \times \tau^{9}}{1.78 \times 0.98)} \times 10 = 0.35\,\text{N}\,\text{cm}^{-1}
$$

[1] $c^2 = a^2 + b^2 - 2ab\cos C$.

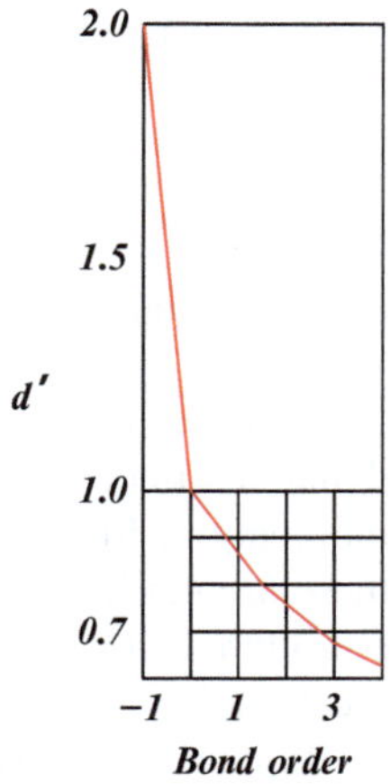

Fig. 5 Variation of interatomic distance with bond order

$$k_\theta = k_r \times 2.19/1.911$$
$$= 0.62\,\text{mdyne/Å rad.}$$
$$R_{\text{vdW}} = 2.76\,\text{Å}; \quad k_r = 0.22\,\text{N cm}^{-1}$$

6.2.2 Interatomic Distance Free of Strain

In principle, these formulae could generate useful parameters for the simulation of a large variety of interactions by force-field methods. For simple diatomic interactions, calculated values of interatomic distance for given bond order can be interpreted as the bond lengths free of strain (ideal bond distances), commonly used in MM. However, for small atoms, such as carbon, in sterically crowded environments, this assumption needs adjustment.

The estimate of interatomic covalent distances proportional to $\tau^n r_0$ assumes spherical charge distributions and does not compensate for distortion by first-neighbor ligands. The effect of such distortion is illustrated schematically below.

Constructive interference between the distorted waves now occurs at a reduced interatomic distance compared to that of the undisturbed waves. Therefore, an interatomic distance free of strain in the correct molecular symmetry environment is reduced to

$$d_0 = d - \tau^{2n} r_0.$$

For first-order C–C interactions, this results in

$$d_0 = 1.545 - 1.78\tau^{10} = 1.531\,\text{Å}.$$

6.2.3 Covalent Interaction

The scheme outlined here has the potential to model all structural and thermodynamic effects, except for torsional flexibility, which depends on orbital angular momentum. Special parameters are needed to model these effects in MM, and these are not included here. Ionic contributions to covalent bonding have not been considered either, but it should also be possible to do so with slight modifications of the model.

Rupture of a covalent bond occurs in stages in the number-theory-based model [38], by transformation into bonds of lower order, until only zero-order interactions (nonbonded) remain. At each stage, a significant rearrangement of the immediate chemical environment accompanies the lowering of bond order. An intimate relationship between all bonds in a molecule is therefore implied. It follows that intramolecular rearrangements, however drastic, do not involve the rupture of bonds. An example is the rearrangement, which involves end-on μ-peroxo-dicopper(II), side-on η^2-peroxo-dicopper(II) and dioxodicopper(III) [39]. In such a process, the molecule preserves its integrity and only experiences a concerted flow of valence electron density, whereby bond orders between all pairs of atoms are affected in a continuous process. The most visible effect could be a modification of the geometric arrangement of nuclei, naturally interpreted in terms of bonds, broken at some points, and re-established elsewhere. This interpretation is certainly not right. Intramolecular cohesion should rather be seen as due to covalent interaction between all pairs of atoms, with non-zero bond orders restricted to neighboring pairs, and fading with increased separation.

In this sense, covalency is a molecular property to be visualized as the interaction within a set of positively charged atom cores in a sea of valence electrons, spreading from local maxima around the nuclei to all of molecular space. The point-charge method models the interaction between all pairs of atoms in the equilibrium configuration. MM does the opposite: from the details of interatomic interactions, it calculates the equilibrium configuration.

6.3 MM Simulation

The heart of MM is a force field [9], and the ultimate force field should be fully transferable between all types of molecules. However, progress towards comprehensive force fields, such as the universal force field (UFF) and derivatives thereof [29, 40], is invariably accompanied by a large increase in the number of parameters. The elegant effort to reformulate MM in terms of valence bond concepts [30] has reduced the number of formal parameters at the expense of generic hybridization parameters. However, the model has not been shown to be applicable for transition metal complexes [33]. These are traditionally modeled by a points-on-a-sphere approach, where the angle function around the metal centers is replaced by nonbonded 1,3-interactions [9, 41, 42], in combination with minor corrections based on ligand-field theory [43] or, in the most general and advanced model for coordination compounds, with a ligand-field (angular overlap model, AOM) term included in the optimization routine [22, 23].

The complexity of problems addressed by molecular mechanics is such that multiparameter modeling is almost unavoidable. The best to hope for is to find a parameter set, based on easily understood chemical concepts. The model outlined above is proposed in that spirit, although considerable refinement is required before it translates into an accurate and therefore useful tool. It is based on the classical concept of electronegativity, reinterpreted in terms of atomic ionization radii and the chemical potential of the valence state. The calculation of these parameters for non-hydrogen atoms does not involve empirical parameters or assumptions. It defines the valence state in terms of characteristic spheres to which a valence electron is confined at uniform charge density. Chemical bond formation occurs on the exchange of this valence charge density between atoms. The consequent polarization, when reduced to point-charge simulation in dimensionless units, serves to describe all covalent dissociation energies as a function of interatomic distance. This function applies specifically to interactions free of strain (the ideal bond distances), which define parameters of fundamental importance in MM. Optimization of a quantized bond-order function allows direct calculation of diatomic dissociation energies and stretching force constants.

The bond-order function applies not only to integral and half-integer bond orders but also to interactions of zero and negative order, characteristic for all nonbonded interactions in a molecule. Based on these ideas, it is in principle possible to define a force field, based on pairwise interactions, which accounts for all structural and thermodynamic effects, apart from those related to orbital and spin angular momenta.

It is not the purpose of this chapter to produce and present a new force field. We rather want to provide a theoretical basis for MM and therefore also to be able to efficiently produce generic force-field parameters. As it stands, one parameter (ionization radius) is needed to initiate the derivation of all other parameters to model all bond orders of any covalent interaction. It is therefore reassuring to note that the uniform valence density within a characteristic atomic sphere has the same symmetry as the 1s hydrogen electron. The first-order covalent interaction between any pair of atoms can therefore be modeled directly by the simple Heitler-London method for hydrogen to predict d, D and k_r [44]. The results are in agreement with those of the simpler number-theory simulation [38], which is therefore preferred for general use.

Table 1 Force-field parameters for alkanes, as obtained by number theory

Parameter	C—C	C—H	H—H	Units
d_0	1.531	1.14		Å
	(1.51)	(1.07)		
k_b	5.12	4.39		$\mathrm{N\,cm^{-1}}$
	(4.88)	(5.09)		
θ_0	106	110	108	deg.
	(106)	(110)	(108)	
$k_\theta(1,3)$	0.64	0.40	0.62	mdyne/Å rad.
	(0.6)	(0.6)	(0.6)	
d_{vdW}	3.56	2.76	1.96	Å
	(3.60)	(2.55)	(2.55)	
k_{vdW}	0.12	0.22	0.40	$\mathrm{N\,cm^{-1}}$
	(0.6)	(0.6)	(0.19)	

Corresponding values obtained by point-charge simulation [8] are shown in *parentheses*

Table 2 Experimental and computed C–C distances for substituted alkanes

Compound	C–C exp. (a) (Å)	C–C cal. (b) (Å)	C–C cal. (c) (Å)	(a)–(c) (Å)
CH_3–CH_3	1.532	1.531	1.531	0.001
CH_3CH_2–CH_3	1.534	1.534	1.533	0.001
$(CH_3)_2CH$–CH_3	1.535	1.537	1.534	0.001
$(CH_3)_3C$–CH_3	1.539	1.541	1.534	0.005
$(CH_3)_3C$–$C(CH_3)_3$	1.582	1.574	1.550	0.032
$((CH_3)_3C)_2$–$CH_3(C(CH_3)_3)$	1.611	1.620	1.566	0.045

(a) Experimental data taken from [9, 14]; (b) calculated using MOMEClite-HyperChem [45, 46] using the MM+ force field [9, 45, 46] with k(C–C) = 4.4 mdyne/Å, r_0(C–C) = 1.523 Å; (c) this work

6.4 *First Results*

As a test for the proposed scheme, the parameters derived for C–C and C–H interactions by point-charge [8] and number-theory simulation were used as a force field for aliphatic hydrocarbons. Note that no optimization of the parameters was attempted, and the results are understood as a feasibility test that obviously needs further refinement. Table 1 summarizes our new parameters derived from number theory together with those from point-charge calculation.

All of the values are in the expected range (see other force fields, such as MM3, Amber and Momec, e.g. in [9, 10, 13, 18, 29, 45]), but by no means are they refined and do not define an accurate force field. The major difference between the number-theory parameters and the alkane force field derived by point-charge simulation occurs in the strain-free C–C bond length with $d_0 = 1.53$ Å compared to 1.51 Å. The optimized structural parameters of a series of aliphatic hydrocarbons, shown in Table 2, although less accurate than with a properly optimized force field, reflect the expected steric variations satisfactorily. Experimental and calculated values with an established empirical force field are shown for comparison [9, 14]. The differences between the sets of calculated bond distances demonstrate the expected similarity between the two force fields and confirm that the new approach may be used to generate a universal force field from first principles. It appears that, in the number-theory-based force field, there is a small imbalance between the attractive bonding forces and the repulsive interactions, and some adjustment might be needed (see also discussion above and note that the nonbonded interactions have been modeled with the simplest possible approach). An obvious and important extension of the method described here, should it be used for the development of a general force field, is the calculation of "electronegativities" for metal ions, in order to also be able to parameterize metal–ligand interactions.

6.5 Discussion

The central idea behind number-theoretic simulation is that chemical interaction happens between reactants in their respective valence states. In the case of diatomic interaction, the valence state is characterized by the ionization radii (electronegativities) of the atoms and is described by simple formulae that relate bond order, interatomic distance and dissociation energy as functions of the golden ratio. It is important to note that the relationships are no more than good approximations for diatomic interaction in polyatomic molecules. For this reason, it is perhaps premature to contemplate a comprehensive data set from which to generate the force-field parameters for the simulation of any classical molecular structure in terms of pairwise interactions. It may even transpire that the predicted diatomic interactions at a many-ligand coordination site are not fully compatible. The simple formulae may require adjustment to be self-consistent. Harmonic force constants would be most sensitive to such modifications. In order to extend the number-theory approach rigorously to complex molecules, it should be necessary to take the chemical environment of interacting atomic pairs into account. As an elementary example, the interaction between two carbon atoms in ethane might be modeled as an interaction between two methyl groups in their molecular valence state. Such second-order corrections may be small and safely disregarded in general applications, but not completely ignored, as experienced in all efforts to construct universal force fields. The feasibility of deriving useful force fields based on number-theoretic diatomic interaction parameters has been demonstrated, but complications that arise on unrestricted extension of the approach are anticipated. The inability to recognize the principles that dictate molecular shape is seen as the most serious constraint. Preliminary ideas, which could serve as an initial guide in the development of algorithms that relate molecular shape to concepts in number theory, are explored next.

7 Molecular Conformation

Both molecular mechanics and wave mechanics are formulated to deal with the intricacies of molecular structure in three-dimensional tangent space. In many cases, where the procedure is clearly inadequate, only minor assumptions are apparently required to remedy small defects. Familiarity with such anomalies eventually conditions the chemist into accepting the ad hoc assumptions as fundamental concepts. The remarkable conviction of most chemists that optical activity only occurs as the collective property of chiral molecules in the bulk is of this kind. It seems to avoid the absurd conclusion that the geometry of a chiral molecule could, by itself, cause optical rotation. Supposedly, it makes more sense to accept that a collection of molecules without S_n symmetry generates the helical motion of charge density that rotates the plane of polarized light. The wave-mechanical identification of molecular magnetic vectors that may interact with polarized light [8] relies on complex variables, but these are routinely eliminated by the separation of spatial variables in quantum-chemical analyses. It is obvious, therefore, that optical activity remains poorly understood. However, it is of more significance that certain aspects of optical activity cannot be rationalized in terms of the wave mechanics of orbital angular momentum [6] and clearly depend on hypercomplex rotation of electronic charge. We identify this observation as one example that demonstrates the four-dimensional nature of molecular conformation.

7.1 *Chirality*

We contend that molecular chirality appears as a four-dimensional symmetry which is incompletely interpreted in three dimensions. The type of anticipated error is demonstrated by the way in which three-dimensional chirality is projected into two dimensions, as in Fig. 6. The two-dimensional chiral system is defined here in the plane that supports the triangular base of a three-dimensional chiral tetrahedron. The symmetry element, shown as a solid vertical line, represents an inversion (I) in three dimensions and a twofold rotation (R) in two. The horizontal broken line represents a twofold rotation in three dimensions, but a reflection (M) in two dimensions. To complete the argument, the three-dimensional reflection that operates diagonally also appears as a two-dimensional reflection. Two-dimensional inversion is equivalent to rotation. In summary,

$$3DI \equiv 2DR$$

$$3DR \equiv 2DM$$

$$3DM \equiv 2DM$$

The two forms on the same side in the frame on the left have the same three-dimensional chirality but different absolute structures in two dimensions. Such a

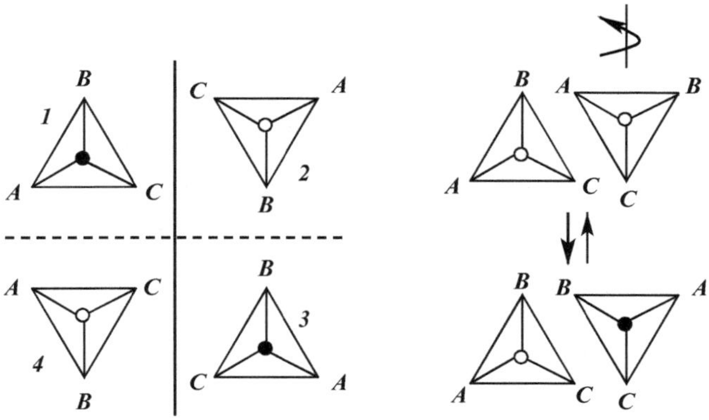

Fig. 6 In two-dimensional projection, the identical pairs of rotated objects 1,4 and 2,3 appear to have opposite chiralities, whereas the enantiomeric pairs 1,2 and 3,4 appear to be identical. The racemization by 3D rotation within a 2D crystal is shown on the *right*

relationship would explain the variable sense of optical rotation, which depends on four-dimensional chirality, in a homochiral family of molecules. A mechanism for sterically unlikely rearrangements is illustrated by the racemization of a two-dimensional chiral crystal. The rearrangement is equivalent to a 2-D reflection that appears to be chemically and sterically forbidden within a crystal, but conveniently achieved by 3D rotation.

7.2 Torsional Interaction

The finer details of molecular shape depend to a large extent on the orientation of fluxional groups on the molecular surface. Whereas MM force fields are based on molecular structures observed in condensed phases, the torsion angles that fix the orientation of such groups cannot be specified. The same probably applies to all torsion angles in the molecular interior. The notorious difficulty of simulating the tertiary structures of large biomolecules in terms of MM pairwise interactions is due to this difficulty to model torsion angles.

In some carefully selected examples, it has been shown how denatured proteins can spontaneously recover their natural folding pattern. The peptide torsion angles, which control the folding, return to the characteristic values of the native protein, known to be independent of chemical factors. Some long-range interaction appears to be at work.

We contend that the shape of large molecules in empty space is affected by the topology of the four-dimensional space–time manifold. Guided by the principle of cosmic self-similarity, it is reasonable to assume that, like many spiral galaxies, extended molecules tend to curve like the surface of a golden spiral. It lies in an elliptic plane, which in four dimensions is the projective space, $\mathbb{S}^3$, with a continuous group structure given by the quaternions, $\alpha + \mathrm{i}\beta + \mathrm{j}\gamma + \mathrm{k}\delta$ with norm $\alpha^2 + \beta^2 + \gamma^2 + \delta^2 = 1$ [47]. On a local scale, the shape of such a molecule is perceived to follow the surface of a large sphere, which is the projection of $\mathbb{S}^3$ into three dimensions. Long-chain molecules will then develop the same spiral structure as kudu horns while two-dimensionally connected macromolecules, such as graphene, will appear spherically distorted. In both cases, the apparent torsion angles of π will more likely appear at a somewhat different value, such as $6/(5\tau^2)$ or $4\sqrt{\tau}$ [6]. Similarly directed torsional modification on an enantiomeric pair would impose different shapes on the molecules and destroy the apparent inversion symmetry that relates them. This minor difference in shape could be responsible for the mysterious homochirality in biological systems.

Acknowledgements The ideas outlined in this chapter took shape during several visits of JCAB, sponsored by the Alexander von Humboldt Foundation, to the Institute of Inorganic Chemistry at the University of Heidelberg. Stimulating interaction with all members of the Institute is gratefully acknowledged.

References

1. Boeyens JCA (2011) In: Putz MV (ed) Quantum frontiers of atoms and molecules. Nova, New York
2. Eddington AS (1948) The nature of the physical world. University Press, Cambridge
3. Batty-Pratt EB, Racey TJ (1980) Int J Theor Phys 19:437
4. Boeyens JCA (2013) Struct Bond 148:71
5. Boeyens JCA, Comba P (2013) Struct Bond 148:12
6. Boeyens JCA (2010) Int J Mol Sci 11:4267
7. Hückel E (1931) Z Phys 70:204. English translation in Hettema H (2000) Quantum chemistry. World Scientific, Singapore
8. Boeyens JCA (2008) Chemistry from first principles. Springer, Dordrecht, Heidelberg, London, New York
9. Comba P, Hambley TW, Martin B (2009) Molecular modeling of inorganic compounds, 3rd completely revised and enlarged edition. Wiley-VCH, Weinheim
10. Burkert U, Allinger NL (1982) Molecular mechanics. ACS Monograph 177
11. Comba P, Kerscher M (2009) Coord Chem Rev 253:564
12. Comba P, Remenyi R (2003) Coord Chem Rev 238–239:9
13. Leach AR (1996) Molecular modeling. Longman, Edinburgh
14. Allinger NL (1977) J Am Chem Soc 99:8127
15. Snow MR, Buckingham DA, Marzilli DA, Sargeson AM (1969) J Chem Soc Chem Commun 891
16. Snow MR (1970) J Am Chem Soc 92:3610
17. Buckingham DA, Maxwell IE, Sargeson AM, Snow MR (1970) J Am Chem Soc 92:3617
18. Clark T (1985) A handbook of computational chemistry. Wiley, New York
19. Niketic SR, Rasmussen K (1977) The consistent force field. Springer, Dordrecht, Heidelberg, London, New York
20. Comba P, Zimmer M (1994) Inorg Chem 33:5368
21. Comba P, Sickmüller AF (1997) Inorg. Chem. 36:4500
22. Burton VJ, Deeth RJ (1995) J Chem Soc Chem Commun 573

23. Deeth RJ (2006) Eur J Inorg Chem 2551
24. Deeth RJ (2007) Inorg Chem 46:4492
25. Deeth RJ, Paget VJ (1997) J Chem Soc Dalton Trans 537
26. Deeth RJ, Munslow IJ, Paget VJ (1997) In: Banci L, Comba P (eds) Molecular modeling and dynamics of bioinorganic systems. Kluwer, Dordrecht, p 77
27. Jansen F, Norrby P-O (2003) Theor Chem Acta 109:1
28. Norrby P-O (2001) In: Cundari TR (ed) Computational organometallic chemistry. Marcel Dekker, New York, p 7
29. Rappé AK, Casewit CJ, Calwell KS, Goddard WA III, Skiff WM (1992) J Am Chem Soc 114:10024
30. Root DM, Landis CR, Cleveland T (1993) J Am Chem Soc 115:4201
31. Cleveland T, Landis CR (1996) J Am Chem Soc 118:6020
32. Firman TK, Landis CR (2001) J Am Chem Soc 123:11728
33. Landis CR, Firman TK, Cleveland T, Root DM (1997) In: Banci L, Comba P (eds) Molecular modeling and dynamics of bioinorganic systems. Kluwer, Dordrecht, p 49
34. Deeth RJ (2011) In: Comba P (ed) Modeling of molecular properties. Wiley-VCH, Weinheim, Germany
35. Meyer L (1870) Liebig's Annalen Suppl. 7:354
36. Boeyens JCA (2008) Z Naturforsch 63b:199
37. Parr RG, Pearson RG (1983) J Am Chem Soc 105:7512
38. Boeyens JCA (2013) Struct Bond 148:93
39. Mirica LM, Ottenwaelder X, Stack TDP (2004) Chem Rev 104:1013
40. Shi S, Yan L, Yang J, Fischer-Shaulsky T, Tatcher J (2003) Comput Chem 24:1059
41. Hambley TW, Hawkins CJ, Palmer JA, Snow MR (1981) Aust J Chem 34:45
42. Bernhardt PV, Comba P (1992) Inorg Chem 31:2638
43. Comba P, Hambley TW, Ströhle M (1995) Helv Chim Acta 78:2042
44. Boeyens JCA (1986) S Afr J Chem 82:361
45. HyperChemTM, Hypercube Inc., 1115 NW 4th Street, Gainesville, Florida 32601, USA
46. Comba P, Hambley TW, Okon N (2001) In: MOMEClite, a molecular modeling package for inorganic compounds; Demo version. Lauer & Okon Chemische Verfahrens- & Softwareentwicklung
47. Stillwell J (1992) Geometry of surfaces. Springer, Dordrecht, Heidelberg, London, New York
48. Putz MV (ed) (2011) Quantum frontiers of atoms and molecules. Nova, New York
49. Hettema H (2000) Quantum chemistry. World Scientific, Singapore
50. Banci L, Comba P (eds) (2001) Molecular modeling and dynamics of bioinorganic systems. Kluwer, Dordrecht
51. Cundari TR (ed) (2001) Computational organometallic chemistry. Marcel Dekker, New York, p 7
52. Comba P (ed) (2011) Modeling of molecular properties. Wiley-VCH, Weinheim, Germany

Struct Bond (2013) 148: 161–179
DOI: 10.1007/978-3-642-31977-8_7
© Springer-Verlag Berlin Heidelberg 2013

All is Number

Jan C.A. Boeyens and Demetrius C. Levendis

Abstract Rational numbers, which correctly describe many recognizable patterns in the physical world, are often seen to converge in the process to irrational limits or even singularities. As a common example, atomic numbers are well known as fundamental parameters in chemistry, but by demonstrating that the periodicity of atomic matter is simulated by the convergence of rational fractions, from unity to the golden ratio, the importance of limiting processes and irrational limits in the modeling of chemical systems and of phenomena such as superconduction is emphasized. Other limiting formulae feature in atomic spectral series, radioactive decay, circular measure, absolute temperature, the speed of light, structure of the solar system and gravitational collapse. In virtually all cases, the convergence involves the irrational golden ratio and the golden spiral, the essential properties of which are briefly reviewed in summary of the arguments developed in this volume. The suspicion that molecular shape should have a related number basis could not be substantiated. Only in the double-helical base pairing of DNA could any correlation between molecular structure and number theory be demonstrated. It is tempting to conjecture that the ubiquitous appearance of irrational limits signals the inadequacy of the $\mathbb{R}^3$ number system to provide a detailed account of the four-dimensional world.

Keywords Discrete set · Golden section · Limiting law · Periodicity · Wave mechanics

---

J.C.A. Boeyens (✉)
Unit for Advanced Scholarship, University of Pretoria, Lynnwood Road, 0002 Pretoria, South Africa
e-mail: jan.boeyens@up.ac.za

D.C. Levendis
Molecular Sciences Institute, School of Chemistry, University of the Witwatersrand, Johannesburg, South Africa
e-mail: demetrius.levendis@wits.ac.za

Contents

1 Introduction

The statement, which inspired the title of this chapter, is traditionally credited to Pythagoras in recognition of the fundamental discovery that the pitch produced by the stretched string on a musical instrument is numerically related to the fractional length of its vibrating part. It was repeated many centuries later in somewhat different terms by Émile de Chancourtois, codiscoverer of elemental periodicity, who claimed [1] that

> ... the properties of the elements are the properties of numbers.

At some stage, most scientists benefit by "the unreasonable effectiveness" of mathematical models in science, but too few look for the cause behind this coinsidence. There is a school of thought that detects a parallel between the subjective choice of topics for analysis and the formulation of matching formulae. At the other extreme, numerical systems are considered to have independent existence in the same way as their physical counterparts. It is, for instance, an undisputed fact that the sophistication of cosmological models faithfully follows developments in number theory. While the concept of infinity remains mathematically unresolved the cosmos remains infinite. However, whether number theories are invented or discovered does not decide their utility in science. If nobody understands the remarkable similarity between Farey sequences, Ford circles and the periodicity of matter, it is no excuse not to exploit this consilience to develop powerful new number-theoretic models for chemistry.

With this prospect in mind, some unexpected parallels between numbers and material behavior will be examined as simplified models of complicated natural phenomena.

2 Numbers and Nature

Most historians trace the rise of modern physics back to Newton and Leibniz who developed the mathematical tools to describe continuous systems and processes. For this reason, the appearance of quantum mechanics, with its emphasis on discrete events, came as a major surprise, still not completely digested. The joke is that Newtonian mechanics concerns the continuous motion of discrete particles whereas quantum theory deals with discrete rearrangements of continuous wave structures. We see a need to differentiate between discrete and continuous systems and to enquire into the feasibility of what happens to be an uneasy compromise, epitomized by the notion of wave–particle duality.

The appearance of numbers in chemistry has an equally interesting history.

The modern view of matter developed from John Dalton's theory which, for the first time, combined the abstractions of atom and element into a single practically useful concept, albeit at the cost of unwanted diversity. Resolution of the dilemma, by postulating hydrogen as the common building block of more complex atoms, was resisted so fiercely that the alleged author, William Prout, had to publish his proposal anonymously. The importance of number was at the heart of his hypothesis. Despite experimental evidence which contradicted the notion, Prout's hypothesis was not without support and remained alive until its final vindication in the discovery of isotopes and atomic number, which ironically also signalled the demise of atomic particle theory.

The next fundamental advance in the theory of matter, like that of Prout, failed to gain recognition, and the name of the author, Hantaro Nagaoka, is now all but forgotten. In line with the old tradition of cosmic symmetry between macrocosm and microcosm Nagaoka conjectured [2] that, like the planet Saturn with its rings, an atom consisted of a solid positively charged core surrounded by a striated electron cloud with regularly spaced local maxima at characteristic distances from the core. In order to avoid electrodynamic instability of the system, the electron density was modeled as standing waves which introduced the set of integers known to characterize atomic spectra. Although never acknowledged by his imitators, Rutherford and Bohr, it is to his credit that he appreciated the defects of a particle model and developed his theory without reference to quantum mechanics.

It is remarkable how the work of William Harkins on the periodicity of matter, dating from the same era, also remained unnoticed and dormant for decades. Whereas Mendeleéff was acclaimed for postulating the properties of undiscovered elements on the basis of his periodic law, Harkins, who characterized seven unknown nuclides on the basis of a more fundamental periodic law, has been forgotten. He was the first to notice, before discovery of the neutron, that the ratio $(A - Z)/A$ for known nuclides never exceeds 0.62 [3], later identified more accurately as the golden ratio, $\tau = 0.61803\ldots$ [4].

The most serious setback for a modern theory of matter was the deliberate suppression of Erwin Schrödinger's demonstration that the behavior of electrons in an atom cannot be described correctly by a particle model and quantum jumps [5, 6]. A beautiful theory, based on a wave model of matter, was buried through professional rivalry to be replaced by incomprehensible concepts such as particles with wavelike properties—even Zitterbewegung, infinite self-energy, probability density, non-Boolean algebra of observables and other weird properties. Remember how Newton described particles as

> solid, massy, hard, impenetrable, movable and indestructible.

Niels Bohr, one of Schrödinger's adversaries, despite his rejection of the wave model, referred on occasion, as quoted by Holton [7], to

> . . . one of the boldest dreams of natural science, namely to build up an understanding of the regularities of nature upon the consideration of pure numbers.

Only waves can generate these numbers.

2.1 Discrete Sets

The Pythagorean theory of music, for lack of a more convincing explanation, became firmly associated with celestial events and structures, to such an extent that initiates like Johannes Kepler could actually hear the "music of the spheres". Although this sound is less audible today, the original theory has survived in the form of the modern standing-wave model of harmonic vibrations. Many other number patterns of the same type have also survived into the modern era as the basis of erudite new theories.

2.1.1 The Bode–Titius Law

This ancient law, which relates the planetary orbits of the solar system to natural numbers, was recently shown [8] to apply to all natural satellite systems in the solar system, subject to a general optimization model which also governs the electronic structure of atoms and the hierarchy of bond orders that characterize covalent interactions in molecules. More than an entertaining exercise in numerology, this law, with the prediction of the asteroid belt to its credit, now emerges as a convincing manifestation of self-similarity on a cosmic scale. An amazing consequence of this symmetry in operation is that the *quantum* effects, hitherto postulated as the distinctive characteristic of microphysical systems, are predicted to be also observable on a cosmic scale. Although not generally endorsed, observations that support this prediction sporadically surface in the recent astronomical literature. It begs a rethink of quantum theory.

2.1.2 Atomic Number

The discoverers of unexpected number patterns are traditionally given a hard time by establishment critics. Nobody more so than William Prout who spotted the tendency of many atomic weights, known in the early nineteenth century, to be improbably close to whole numbers. He was ridiculed and maligned almost as badly as Alfred Wegener, who first mooted the idea of continental drift. Wegener never received the recognition that he deserved, arguably for overlooking the "true" basis of the phenomenon. The same happened to Prout and his handful of supporters. In order to account for exceptional atomic weights, like the 35.5 of chlorine, there are several accounts on record, proposing the possibility of such elements being mixtures of atoms that differ in mass but not in chemical properties. William Harkins who actually implicated the existence of a *neutron*, many years before the discovery of such a particle, also failed to be taken seriously. What this history shows, is that the idea of number led the way for about a century. It ceased to be a joke only when imagination was no longer needed in order to appreciate the obvious.

This saga has still not ended. Once the penny had dropped and the many-headed finally accepted the notion of isotopes, their thinking stagnated again into a particle count to explain the appearance of integers. This is the brick wall that stopped Schrödinger when he proposed a wave model as the natural origin of quantum integers.

2.1.3 The Balmer Formula

The quantum theory of matter developed from an innovative suggestion by Johann Balmer to account for the optical spectrum of atomic hydrogen, observed as a series of emission lines at narrowly defined wavelengths. He related the measured wavelengths to a series of integers by the formula

$$\frac{1}{\lambda} = R\left(\frac{1}{4} - \frac{1}{n^2}\right), \qquad n = 3, 4, 5 \ldots,$$

that existed as an unexplained oddity outside of mainstream physics for many years, until a matching relationship with the extranuclear distribution of electrons on an atom was conjectured by Nagaoka and later incorporated into the Bohr formula, which linked it to the quantum theory of radiation. It is interesting to note that both Nagaoka and Bohr based their proposals on a perceived likeness between atoms and celestial structures.

The self-similarity between atoms and planetary systems goes even further. Whereas an atomic nucleus is surrounded by an electronic core and a valence shell, the solar system has an inner core of rocky planets with major planets beyond the asteroid belt. A major planet like Saturn has a core of rings and minor shepherding moons, surrounded by regularly spaced major moons. In the case of the solar system, all planets have characteristic quantum numbers that converge from both ends to the golden ratio at the chaotic asteroid belt. The boundary between planetary rings and major moons is likewise defined by τ.

By reference to Fig. 2, which follows later on, a chaotic atomic core is readily identified as the regions labelled s, d and f. It is precisely for the regular p-block elements that the covalence formulae based on the golden ratio work best.

2.2 Irrational Limits

The Balmer formula defines an infinite series that converges to the limit in which $\lambda = 4/R$. On approaching this limit, the spectral lines are bunched together more closely, such that the discrete line spectrum turns into a continuum. The spectrum becomes blurred, but the limit is accurately measurable.

On closer examination, the atomic numbers of stable nuclides, consisting of Z protons and N neutrons, are also seen to converge to $Z = 102$ as the ratio of Z/N approaches a constant value, now recognized as the golden ratio, τ. Remarkably, the same constant characterizes the logarithmic spiral that generates the Bode–Titius law. It is therefore of interest to examine other well-known limiting processes which could reveal a more general common pattern.

2.2.1 Radioactivity

Radioactive decay is one example of a natural process, said to proceed *exponentially* as a function of time. In practice, it is observed that the measured activity A at time t has diminished to half its initially measured value of A_0 after a characteristic time interval known as the half-life of the process. This result is valid for any A_0. In explanation, the rate process of radioactive decay has been formulated as

$$A = A_0 \mathrm{e}^{-\lambda t} ,$$

according to which

$$\frac{A_0}{2} = A_0 \mathrm{e}^{-\lambda T_{\frac{1}{2}}}$$

for half-life $T_{\frac{1}{2}}$ and decay constant λ. In logarithmic notation the expression

$$2\mathrm{e}^{-\lambda T_{\frac{1}{2}}} = 1$$

reduces to $\ln 2 = \lambda T_{\frac{1}{2}}$, which defines constant $T_{\frac{1}{2}}$.

This result demonstrates the important principle that, in theory, the decay process can continue forever, albeit with diminishing activity. At any given time, another period of $T_{\frac{1}{2}}$ is required to half the currently measured activity. Theoretically a nearly infinite number of half-lives expire before all activities cease. The decay process may therefore be considered as approaching a final unreachable state ever more closely.

Mathematically the process is described precisely by the exponential function, which is based on e, the basis of natural logarithms. This constant is presented to the first 3,441 decimal places on a coffee mug for sale in Germany. It also defines e by the formulae

$$\mathrm{e} = \lim_{n\to\infty}\left(1+\frac{1}{n}\right)^n = \sum_{k=0}^{\infty}\frac{1}{k!} \quad , \quad \mathrm{e}^{\mathrm{i}\pi} = -1 \quad \text{and} \quad \mathrm{e}^{\mathrm{i}\varphi} = \cos\varphi + \mathrm{i}\sin\varphi .$$

None of these can ever specify the ultimate value of e, and any refinement to improve the accuracy can be continued indefinitely, without the possibility of ever reaching finality.

The remarkable fact is that the exponential function had been known for centuries before the discovery of radioactivity and its formulation could therefore not have been inspired by the natural process. One may ask, what led Euler to this invention—or discovery?

2.2.2 Circular Measure

Another well-known transcendental number, called π, measures the ratio between the diameter and circumference of a circle in the Euclidean (or complex) plane. Like e, it is associated with a well-defined but unreachable limit—in this case the circumference of a regular polygon with infinitely many sides. For regular polygons with an even number of sides, it is easy to demonstrate that the ratio between longest diameter and total circumference is calculated as $\pi_N = N \sin(180°/N)$. This value increases steadily with N from $\pi_4 = 2.83$, $\pi_6 = 3$, $\pi_8 = 3.06$, $\pi_{64} = 3.1403$ to $\pi_{1,024} = 3.1415877$ and beyond. It approaches but never reaches the value of π, which, like e, is an irrational number that keeps many computers occupied to specify its value with improving accuracy. A second coffee mug defines

$$\pi = 4 \times \left(1 - \frac{1}{3} + \frac{1}{5} - \frac{1}{7} + \frac{1}{9} - \frac{1}{11} + \frac{1}{13} - \frac{1}{15} + \frac{1}{17} - \frac{1}{19} + \cdots\right),$$

which oscillates convergently around $\pi = 3.141592\ldots$.

There is an intriguing philosophical issue related to such a pointless search for infinity. The elusive absolute value of π, and perhaps of e as well, has a different value in curved space and a completely different meaning in curved space–time. The concept of a linear infinite measure has, for example, no meaning in topologically closed spaces. The limiting point that quantifies π and e only exists in hypothetical Euclidean space, which provides a convenient frame of reference for projection from the globally curved underlying space–time into three-dimensional local tangent space. Humans are conditioned to analyse all observations in familiar three-dimensional tangent space as a function of universal time.

2.2.3 Zero Temperature

Probably the most familiar example of an unreachable limit in physical science is the absolute zero of temperature. Rather than a mathematical impossibility, 0 K implies the cessation of all motion, which renders it experimentally inaccessible. A logical explanation of the effect may well be another futile search for infinity. Energy, matter and motion can only occur in curved space–time. An eventless situation is, by definition, restricted to flat, infinite Euclidean space, which occurs nowhere in the real world. In this sense, the third law of thermodynamics and phenomena such as superfluidity, which occurs near 0 K, provide experimental evidence of space–time curvature.

2.2.4 The Speed of Light

The most famous limitation in physical science is the limiting speed c, imposed by the special theory of relativity and known as the speed of light. Again, this is a limitation imposed by the mathematical properties of Euclidean space on the perception of unaccelerated relative speed as measured in different frames of reference. The inability to measure the relative speed of electromagnetic signals is perceived as distortion of space–time geometry in the form of spacial contraction, time dilation and increasing mass. All of these perceptions attest to the non-Euclidean nature of four-dimensional events. The familiar dynamical concepts of classical mechanics, as mere projections, have no meaning in four-dimensional space–time. Classical dynamics refers to the analysis of observations as a function of three-dimensional Euclidean space and one-dimensional universal time. The projection of four-dimensional phenomena into this tangent space creates the impression of magic at work.

2.2.5 The Black Hole

The mother of all irrational limits is the singularity at the center of a black hole. It epitomizes the infinite gravitational field at $r = 0$, predicted by the inverse-square law of force, $F = -GMm/r^2$, and its effect on a run-away accumulation of matter. Despite many erudite theories, it remains the foremost enigma of astronomy and the harshest test for the infinity concept. A proposed resolution of the problem [9] seeks an outlet for the infinitely accumulated matter through an Einstein–Rosen bridge across the interface that separates antipodal regions of the double cover of the projective space–time model. We consider the conjecture that the universe originated in an initial singularity to be completely over the top.

2.2.6 The Common Limit

Many limiting processes in nature appear to be associated with the irrational numbers π, e and τ, each of which is defined geometrically in three-dimensional tangent space. The geometry of a circle defines π, a regular pentagon defines τ and e relates to the complex plane.

These definitions are no longer valid in 4D curved space–time. It has been known since 1603 by the theorem of Thomas Harriot [10] that for a spherical triangle Δ, with angles α, β and γ, the angular excess

$$\delta = \alpha + \beta + \gamma - \pi$$

is proportional to the area(Δ), and hence non-zero.

Alternatively, the radius of a great circle, constructed in the surface of a unit sphere, $r = 1$, define an elliptic $\pi_e = 2$. Without further proof, it becomes obvious that although the excess in curved space–time may not be measurable in local tangent space, it necessitates serious correction over astronomical distances. Extrapolation to the subatomic scale is equally hazardous, with the effect of 4D curvature no longer a negligible factor. In a sense, each irrational limit in tangent space amounts to a three-dimensional singularity, which disappears in curved space–time. The Pythagorean cult disintegrated for a similar reason following the discovery of irrational numbers, a concept at variance with their dogma. If modern scientists had the same courage of their convictions, singularities and infinities in physical theory would be anathema. Instead we disguise our ignorance as subtle theory.

Singularities and infinities are only unreachable on analysis in an inadequate number system. The irrational limits that emerge in three-dimensional space are regular points in four-dimensional space–time. By following natural numbers as they appear in three-dimensional theory of chemical systems invariably leads to irrational numbers, such as the golden ratio, which signals progression to four dimensions.

Resolution of the singularity problem in four dimensions can only be guessed at. A possible guide towards a mathematical elucidation is provided by the mapping of a spiral inside a golden rectangle. It amounts to inscribing semicircles within the squares that divide the rectangle into smaller copies of itself. The process continues as an endless approach towards the unreachable focus of the spiral at the intersection of rectangle diameters. The same process unfolds in the opposite direction on addition of larger squares to the growing golden rectangles. Instead of the infinitely small, the spiral now extends to the infinitely large. In four-dimensional projective space, the two points of convergence may be identified as a single point at infinity.

On this basis, we conjecture that the golden spiral exists in 4D as two conjugate parts of a closed construct, which also models the natural number system.[1] The imaginary half represents extension of reality into a parallel antiworld. Singular points in gravity, temperature, the composition of atomic nuclei and the number system, are now identified as the 4D points of contact between conjugate worlds. There is no closer approach to 4D reality and no better device to model this reality. It is no longer surprising to find that chemical systems are amenable to golden spiral optimization and easier to understand why chemical interactions are accurately modeled by the resulting wave structure.

[1] The geometrical definition of both π and τ depends on space–time curvature, and since both e and τ represent converging number sequences, the identity $e^{i\pi} = -1$ cannot hold universally, unless the number system also adapts.

2.3 The Golden Section

The irrational number, known as the golden ratio, is said to be the most irrational of them all. Like other irrationals, it also occurs as the limit of a regular series of rational fractions, in this case the Fibonacci fractions. In nature, it occurs as the convergence limit of the mass fractions of stable nuclides, $Z/(A-Z)$. As a clue to its physical meaning, it is noted that the stability of nuclides depends on their space–time environment [4]. In regions where space–time curvature approaches infinity, the mass ratio $Z/(A-Z) \rightarrow 1$. In the hypothetical situation of zero curvature, matter does not exist. It is inferred that in an intermediate situation of curvature, conducive to the development of biological life, the mass ratio $Z/(A-Z) \rightarrow \tau$.

Some of the remarkable properties of the golden section, with relevance to chemistry, are summarized in the introductory chapter of this volume. Perhaps most surprising is the close relationship between golden ratio and the Fibonacci series, which was first formulated to model the population growth in a rabbit colony.

Not only the golden ratio itself but also its integral powers, τ^n, are embedded as convergent series in the Fibonacci sequence. Comparison of the first few of these reveals a striking pattern:

$$
\begin{array}{lccccccccll}
 & & & & & & & & & & m \\
\tau : & \frac{1}{1} & \frac{1}{2} & \frac{2}{3} & \frac{3}{5} & \frac{5}{8} & \frac{8}{13} & \frac{13}{21} & \frac{21}{34} & \rightarrow 0.6180\ldots & 1 \\
\tau^2 : & \frac{1}{2} & \frac{1}{3} & \frac{2}{5} & \frac{3}{8} & \frac{5}{13} & \frac{8}{21} & \frac{13}{34} & \frac{21}{55} & \rightarrow 0.3819\ldots & 1 \\
\tau^3 : & \frac{1}{3} & \frac{1}{5} & \frac{2}{8} & \frac{3}{13} & \frac{5}{21} & \frac{8}{34} & \frac{13}{55} & \frac{21}{89} & \rightarrow 0.2360\ldots & 2 \\
\tau^4 : & \frac{1}{5} & \frac{1}{8} & \frac{2}{13} & \frac{3}{21} & \frac{5}{34} & \frac{8}{55} & \frac{13}{89} & \frac{21}{144} & \rightarrow 0.1458\ldots & 3 \\
\tau^5 : & \frac{1}{8} & \frac{1}{13} & \frac{2}{21} & \frac{3}{34} & \frac{5}{55} & \frac{8}{89} & \frac{13}{144} & \frac{21}{233} & \rightarrow 0.0901\ldots & 5 \\
\tau^6 : & \frac{1}{13} & \frac{1}{21} & \frac{2}{34} & \frac{3}{55} & \frac{5}{89} & \frac{8}{144} & \frac{13}{233} & \frac{21}{377} & \rightarrow 0.0557\ldots & 8
\end{array}
$$

$$\tau^n = \lim_{i\rightarrow\infty} \frac{F_i}{F_{i+n}}$$

The modularity of each sequence, defined by the matrix of successive fractions, for the series τ^n,

$$m = \begin{vmatrix} n_i & n_{i+1} \\ d_i & d_{i+1} \end{vmatrix} = \pm F_n \, ,$$

the nth Fibonacci number.

The unique properties of the golden section show that the τ-sequence

$$F_\tau = \ldots, \tau^n, \tau^{n-1}, \ldots, \tau^0, \tau^{-1}, \tau^{-2}, \ldots, \tau^{-n} \ldots$$

is not only a double geometric sequence but also a Fibonacci sequence. The generalized Fibonacci series, $G^{8,2}$, also known as the Funk-Hellet sequence [11], with seed values $S_0 = 8,\ S_1 = 2$, reduces to $1{,}000F_\tau$, rounded off, on aligning G_0 with τ^{12}:

n	1	2	3	4	5	6	7	8	9	10	11	12	13	14
F_τ	τ^{11}	τ^{10}	.	.	.	.	.	.	.	.	.	τ^0	τ^{-1}	τ^{-2}
$1{,}000F_\tau$	5	8	13	21	34	56	90	146	236	382	618	1,000	1,618	2,618
$G^{8,2}$	2	10	12	22	34	56	-	-	-	-	-	-	-	-

Note the special properties of $G_{14} = 2 \times 7 \times 11 \times 17$, $12G_{14} = 10^4\pi$.

The construction of a golden spiral is also readily reduced to Fibonacci numbers. A dilative rotation of 2π transforms any point on the spiral into a homothetic point, which is similarly placed and directed. The spiral, said to be homothetic to itself, therefore, has the property of self-similarity at all scales. This means that the spiral is endless in both directions. When seen to converge towards a focal point, where the diagonals of the golden rectangles intersect, it means that the spiral continues in more dimensions within a cone. The diverging segments meet at a point at infinity to create a single closed curve in projective space. This could be the property referred to by Johannes Kepler in stating that,

> ... the divine proportion served the Creator as an idea when He introduced the continuous generation of similar objects from similar objects.

Logarithmic spirals are widely used, especially in the engineering sciences and for botanical phyllotaxis to solve optimization problems. In the case of a distribution with multiple sub-maxima, the procedure consists in the variation of a divergence angle between adjacent maxima on the spiral until the optimal distribution is realized.

2.3.1 Superconduction

Metallic superconductors that operate at temperatures below 30 K were discovered a hundred years ago, followed by high-temperature superconductors ($T_c \sim 90$ K) 75 years later. A centennial survey [12] concluded:

> Exactly how such high-temperature superconductors (HTSS) work remains a mystery, but that has not stopped engineers from trying to exploit them.

However, a plausible mechanism, based on the golden ratio and which applies to all superconductors, exists [13] as the property of atomic nuclei rather than the interaction of electrons with lattice phonons.

The observation that the convergence of $Z/N \rightarrow \tau$ promotes nuclear stability was interpreted to imply a packing mode for nucleons in three-dimensional analogy of botanical phyllotaxis. The general property, $N > Z$, of stable nuclides then results in a golden excess of $x = Z - \tau N$ protons, which form an electrically

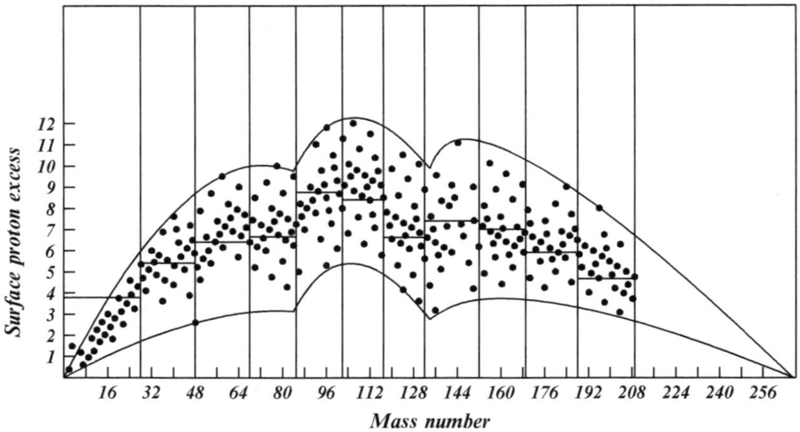

Fig. 1 Proton surface excess, $x = Z - \tau N$, as a function of mass number. Nuclide periodicity predicts maximal surface spin to occur in the regions as marked, in general agreement with the measured spin and elemental superconductivity of odd mass number nuclides

positive layer in the nuclear surface, according to Fig. 1. The analysis predicts enhanced superconduction of isotopically pure materials in favourable allotropic modification, such as low-temperature ^{117}Sn with spin of 11/2. These predictions await technological exploitation.

The existence of high-spin surface layers suggests an intriguing new interpretation of magnetic phenomena associated with superconduction. To account for the anisotropy and anomalous signs of Hall coefficients, it is noted that an external magnetic field simply aligns the excess spin system to generate an oppositely directed internal magnetic field. Alternatively the internally generated magnetic field of a superconductor is sufficient for the diagnostic levitation of a small magnet over the system, in what is known as the Meisner effect.

It is important to note that whereas the BCS theory only applies to special low T_c materials, the golden-excess model covers all superconductors. The disadvantage of currently known HTSS is the brittleness of the ceramic materials that complicates the manufacture of flexible wires made up of well-aligned crystals. Raising the critical temperature, T_c, of cheap industrially important metals by isotopic enrichment offers a more manageable alternative technology.

3 Chemistry by Number

The most important contribution of chemistry to science, the periodic table of the elements, can be derived directly by elementary number theory. The fundamental relationship between Farey sequences and Ford circles is readily demonstrated.

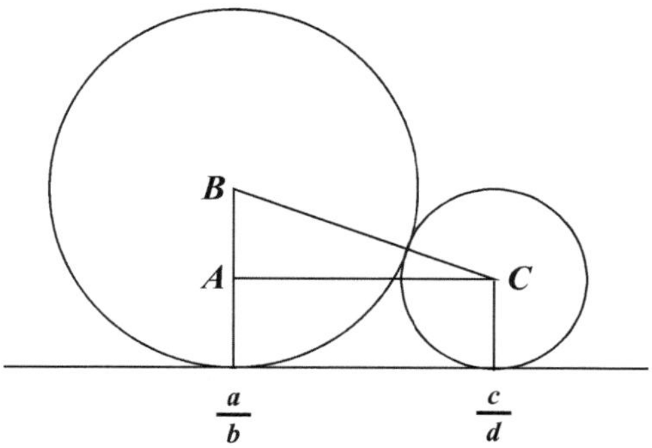

Consider two touching circles with diameters of $1/b^2$ and $1/d^2$, centered at respective points a/b and c/d as shown in the diagram, and such that $b < d$ and $a/b < c/d$. By construction, in the right triangle ABC:

$$AB = \frac{1}{2b^2} - \frac{1}{2d^2}$$

$$AC = \frac{c}{d} - \frac{a}{b}$$

$$BC = \frac{1}{2b^2} + \frac{1}{2d^2} \quad \text{(sum of radii)}$$

By Pythagoras:

$$(BC)^2 - (AB)^2 = \frac{1}{2(bd)^2} + \frac{1}{2(bd)^2} = \left(\frac{1}{bd}\right)^2 = (AC)^2 = \left(\frac{bc - ad}{db}\right)^2$$

Hence,

$$bc - ad = \begin{vmatrix} c & a \\ d & b \end{vmatrix} = 1$$

This is the unimodular condition that characterizes neighboring fractions in a Farey sequence.

Ford-circle mapping of the periodic table [13], based on this idea, consists of a central circle flanked on both sides by three satellites, which together represent the $\mathscr{F}_4$ Farey sequence:

$$\mathscr{F}_4 = \left\{ \frac{0}{1}\ \frac{1}{4}\ \frac{1}{3}\ \frac{1}{2}\ \frac{2}{3}\ \frac{3}{4}\ \frac{1}{1} \right\}$$

In each fraction, h/k, the denominator is interpreted to distinguish between periodic systems, $k = 1 \to 4$, of size $2k^2$. The different fractions in $\mathscr{F}_4$ represent subsets with $j = 2\sum_{i=1}^{k}(2i-1)$, i.e. $2(1+3+\cdots \leq k)$, elements. Each subset with fixed k expands into a sequence of smaller fractions, n/j, $n = 0 \to j$. On arranging these fractions symmetrically around $\frac{1}{2}$, the layout of the periodic table [14] is reproduced, as shown in Fig. 2, providing each block represents two chemical elements in the order defined by the arrows.

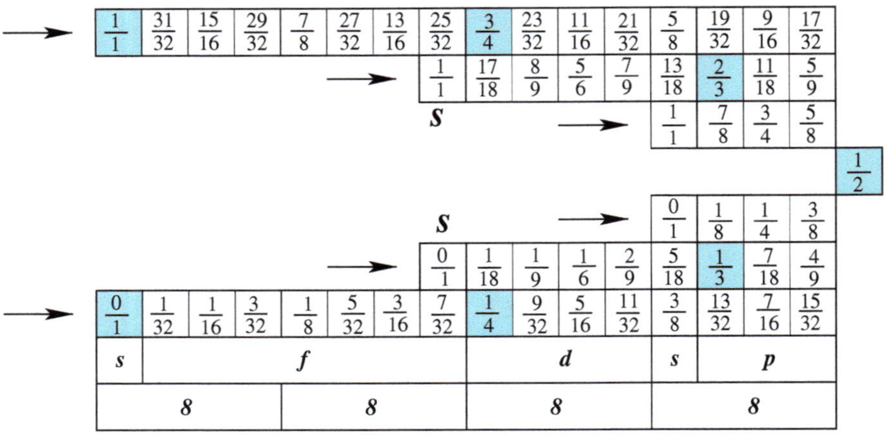

Fig. 2 Periodic table of the elements, defined directly by the Farey sequence $\mathscr{F}_4$. Shaded blocks show how the Farey fractions subdivide the periods into regions that correspond to the s, p, d, f blocks of chemical elements. Blocks of two and eight elements define the compact form of the appended periodic table

Without assuming an exclusion principle, the Farey fractions subdivide the periodic sets into subsets that correspond to the traditional s, p, d, f energy levels, derived from atomic spectra. The resulting sublevel structure is highlighted by shading in Fig. 2. It is necessary to emphasize that these fractions do not refer to Z/N ratios.

Apart from the relative abundance of isotopes, all other basic data usually shown on a periodic table of the elements are predicted by elementary number theory. Based on the periodic table shown as an Appendix, the chemically important parameters of ionization radius and electronegativity have been used, together with the golden section and golden spirals, to derive all essential parameters pertaining to covalent interaction and the optimization of molecular structure by MM. The detailed results are described in the preceding chapters in this volume.

3.1 Molecular Structure

Chemical science is based on more than atomic structure. More precisely, it concerns the properties of molecules, and this is where all current theories fall short. Efforts to simulate molecular structure by elementary number theory have been equally unsuccessful, with the exception of one partial success pertaining to the structure of DNA base pairing.

A mysterious number code, known as *Ho Tu* [15], emerged in China during the fourth millennium BCE, in the form of 55 dots arranged as in Fig. 3. The circled clusters represent *yin-yang* trigrams in the eight radial compass directions, arranged in an 8 × 8 matrix, known as *I Ching*, generate the numbers 0–63 as hexagrams in binary code.

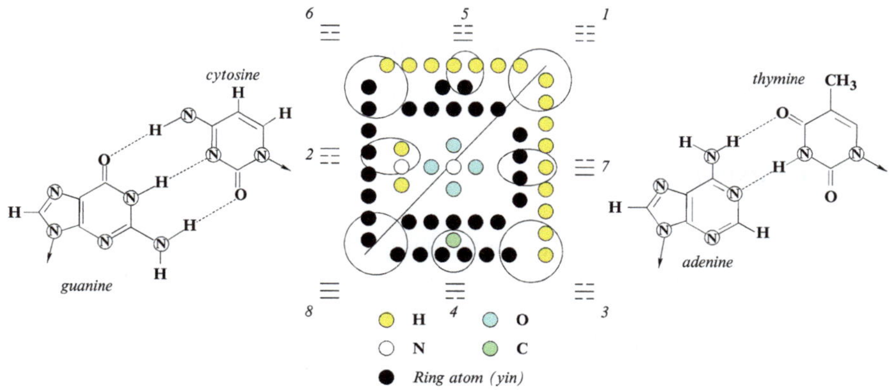

Fig. 3 Ho Tu simulation of DNA base pairing

Not only do the I Ching hexagrams correspond to the DNA genetic code [16], but the Ho Tu dots have also been shown to symbolize the molecular structure of the two H-bonded base pairs. The two base pairs are made up of 55 atoms, just like the Ho Tu consists of 55 dots. The dots above the diagonal in Fig. 3 model the cytosine–guanine pair, and those below the diagonal represent adenine–thymine, as drawn.

In each Ho Tu segment, there are 15 black (yin) dots, just as each base pair has 15 central ring atoms. These 15 atoms define a (5+4)- and a 6-membered cyclic system. The five central yang dots allude to the five hydrogen bonds in groups of 3 and 2. The extracyclic H, N, O and C atoms are colour coded to emphasize the correlation. The match between Ho Tu and the base pairing is complete.

The equivalence of I Ching and the binary number system is beyond dispute. Its correspondence with the 64-codon DNA genetic code [16] is equally convincing. Derivation of the 8-trigram compass mandala, which generates the I Ching from the Ho Tu pattern, is readily demonstrated. Unless the homology of the Ho Tu and DNA base pairing is a massive coincidence, there can be little doubt that it presents a detailed code of biological replication still not fully understood. The origin of the code is shrouded in the mists of time. Legend has it that the code was delivered to the emperor Fu Hsi by a "dragon horse", which has been likened to a rocket from outer space.

The cross-sectional profile of the most common form of crystalline DNA, with a 10_1 screw axis, has an enclosing form scaled by the limiting factor [17] of

$$\mu(10/3) = \frac{\cos 3\pi/10}{\cos \pi/10} = \tau\,,$$

the golden ratio. This form of DNA double helix also represents the most efficient space-filling arrangement.

The degree of success in relating the structure of the most intricate of molecules to pure numbers raises the haunting suspicion that the inability to formulate a valid theory of molecular structure may be traced back to an inadequate understanding of cosmic self-similarity. With a little insight, the accumulated empirical wisdom on molecular conformation could perhaps fall into place to solve the mystery of molecular shape.

4 Conclusion

In all cases where the golden section or the golden spiral correlates with chemical phenomena, convergence to some singularity is observed. The most striking example, shown in Fig. 4, occurs as the composition of stable nuclides, measured as Z/N, converges to the golden ratio as $Z \to 102$. At the same time, the hem lines, which define nuclide periodicity of 24, map out the observed periodic table of the elements at $Z/N = \tau$.

Extrapolation of the hem lines from 0 to 1.04 defines special periodic functions at $Z/N = 0, 0.58, \tau$, 1 and 1.04, with $0 \leq Z \leq 102$. The periodic function defined at $Z/N = 0$ refers to neutron periodicity; at $Z/N = 0.58$, the sequence of points corresponds to the Schrödinger solution for H; the periodic functions at $Z/N = 1$ and 1.04 are inverse to those at τ and 0.58, respectively. The logic behind this is that the Schrödinger model, which ignores all but the Coulomb attraction between proton and electron, refers to an empty (Euclidean flat) universe, inverse to the infinitely curved geometry at $Z/N = 1.04$. As the general curvature of space–time follows a golden spiral, the observed periodicity occurs at $Z/N = \tau$ and nuclear synthesis by α-particle addition at $Z/N = 1$. The structure inversion is due to the known effect of pressure on electronic energy levels in atoms.

By the same argument, the ratio of $Z/N = \epsilon \simeq 0.58$ should be interpreted as characteristic of free space. It is of interest that also the ubiquitous mysterious ratio [18] of $0.1115 \simeq 2\epsilon$. Maybe it is no accident that the approximate formula for a Fibonacci spiral [19]

$$r \simeq 2\epsilon \exp(\theta \cdot \tau/2)$$

contains both ϵ and τ, in line with the conjecture that the golden spiral embodies the relationship between Euclidean tangent space and the underlying four-dimensional curved space–time manifold.

The points at $Z/N = 1.04$ are arranged symmetrically around a central point. On reflection of the entire pattern around the line at $Z = 51$, a set of closed lines, which cannot be embedded in two dimensions, is generated. Drawn on the surface of a Möbius band, the double pattern is readily identified as representing both matter and antimatter and requires the points at 0 and 102 to coincide. In order to achieve this, the single edge of the Möbius band must be glued to itself to create a closed space which can no longer be embedded in three dimensions and is known as projective space.

We conclude that all of the results described in this volume can be generated from two related constructs—the golden spiral and projective geometry. Remarkably, models of exactly these two things have recently appeared on the science campus Im Neuenheimer Feld of the Ruprecht-Karls University of Heidelberg where a large part of this work was done. These are pictured with one of us in Fig. 5.

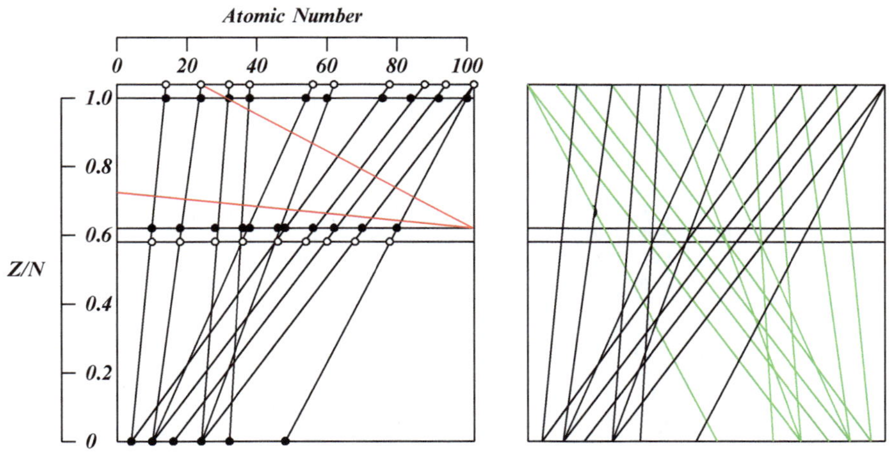

Fig. 4 Variability of the periodic table of the elements, a subset of the 24-member periodicity of stable nuclides, depends on space–time curvature and is mapped by 11 hem lines in the frame on the *left*. The field of stability is defined by the limiting lines that converge to τ at $Z = 102$. The symmetrical version on the *right* is only resolved in 4D projective space

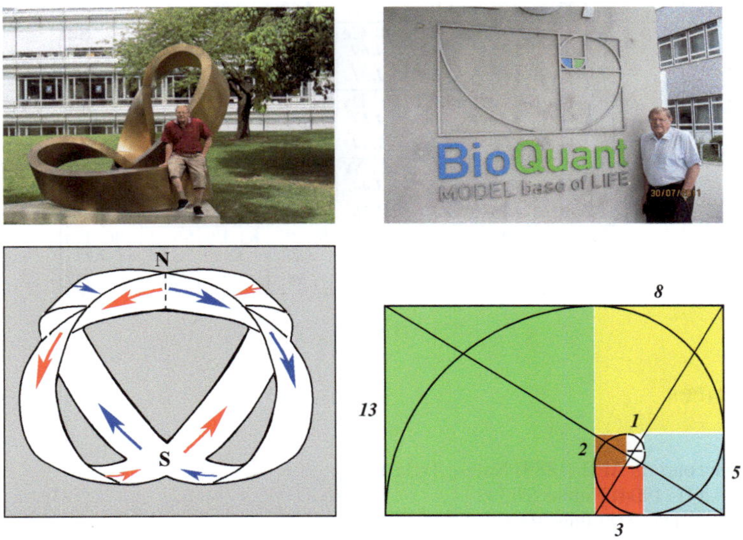

Fig. 5 Projective space is visualized by the space-filling intersection of an infinitude of infinitesimal Möbius strips, like the two bands, shown in the *lower left frame* [8]. A golden spiral results from inscribing circular segments into the gnomons defined by Fibonacci numbers in a *golden rectangle*

Appendix: Periodic Table of the Elements

2	1 H 0.98 6.25	2 He	Z Sym r_0(Å) χ	Closed s shell	Closed p shell	Closed d shell	CLosed f shell	Radio–active
8	3 Li 2.36 2.60	4 Be 2.20 2.79	5 B 1.88 3.26	6 C 1.78 3.45	7 N 1.69 3.63	8 O 1.60 3.83	9 F 1.52 4.03	10 Ne 1.44 4.26
8	11 Na 3.09 1.98	12 Mg 2.87 2.14	13 Al 2.74 2.24	14 Si 2.62 2.34	15 P 2.51 2.44	16 S 2.47 2.48	17 Cl 2.30 2.67	18 Ar 2.21 2.78
18	19 K 3.50 1.75	20 Ca 3.08 1.99						
	21 Sc 3.02 2.03	22 Ti 2.99 2.05	23 V 2.96 2.07	24 Cr 3.02 2.03	25 Mn 2.90 2.11	26 Fe 2.87 2.14	27 Co 2.85 2.15	28 Ni 2.81 2.18
	29 Cu 2.88 2.13	30 Zn 2.76 2.22	31 Ga 3.00 2.04	32 Ge 2.89 2.12	33 As 2.80 2.19	34 Se 2.71 2.26	35 Br 2.61 2.35	36 Kr 2.52 2.43
18	37 Rb 3.81 1.61	38 Sr 3.54 1.73						
	39 Y 3.27 1.88	40 Zr 3.24 1.89	41 Nb 3.30 1.86	42 Mo 3.26 1.88	43 Tc 3.14 1.95	44 Ru 3.20 1.92	45 Rh 3.16 1.94	46 Pd 2.29 2.68
	47 Ag 3.11 1.97	48 Cd 3.00 2.04	49 In 3.31 1.85	50 Sn 3.19 1.92	51 Sb 3.09 1.98	52 Te 2.98 2.06	53 I 2.88 2.13	54 Xe 2.75 2.23
32	55 Cs 4.03 1.52	56 Ba 3.75 1.64	57 La 4.62 1.33	58 Ce 4.69 1.31	59 Pr 4.68 1.31	60 Nd 4.66 1.32	61 Pm 4.64 1.32	62 Sm 4.63 1.32
	63 Eu 4.61 1.33	64 Gd 4.51 1.36	65 Tb 4.58 1.34	66 Dy 4.57 1.34	67 Ho 4.55 1.35	68 Er 4.54 1.35	69 Tm 4.52 1.36	70 Yb 4.50 1.36
	71 Lu 3.54 1.73	72 Hf 3.50 1.75	73 Ta 3.47 1.77	74 W 3.44 1.78	75 Re 3.40 1.80	76 Os 3.37 1.82	77 Ir 3.34 1.84	78 Pt 3.41 1.80
	79 Au 3.38 1.81	80 Hg 3.24 1.89	81 Tl 3.43 1.79	82 Pb 3.32 1.85	83 Bi 3.22 1.90	84 Po 3.12 1.97	85 At 3.03 2.02	86 Rn 2.93 2.09

References

1. de Chancourtois AEB (1889) Nature 41:186
2. Nagaoka H (1904) Nature 69:392
3. Harkins WD (1931) Phys Rev 38:1270
4. Boeyens JCA (2003) J Radioanal Nucl Chem 259:33
5. Schrödinger E (1926) Naturwissenschaften 28:664
6. Schrödinger E (1952) Br J Phil Sci 3:109
7. Holton G (1952) Introduction to concepts and theories in physical science. Addison-Wesley, Reading
8. Boeyens JCA (2009) Phys Essays, 22:493
9. Boeyens JCA (2010) Chemical cosmology. Springer, Dordrecht, Heidelberg, London, New York

10. Stillwell J (1992) Geometry of surfaces. Springer, Dordrecht, Heidelberg, London, New York
11. Funk-Hellet C (1956) La Bible et la Grand Pyramide d'Egypte. Hellet Vincent, Montreal
12. The Economist Technology Quarterly, December 3rd 2011: Superconductors 100 years on
13. Boeyens JCA, Levendis DC (2008) Number theory and the periodicity of matter. Springer, Dordrecht, Heidelberg, London, New York
14. Boeyens JCA, Comba P (2013) Struct Bond 148:1
15. Walter K (1994) Tao of chaos. Kairos Center, Austin
16. Boeyens JCA (2003) Cryst Eng 6:167
17. Janner A (2001) Cryst Eng 4:119
18. Maddox J (1983) Nature 304:11
19. Boeyens JCA (2013) Struct Bond 148:71

Index

MIX
Papier aus verantwortungsvollen Quellen
Paper from responsible sources
FSC® C105338

If you have any concerns about our products,
you can contact us on
ProductSafety@springernature.com

In case Publisher is established outside the EU,
the EU authorized representative is:
Springer Nature Customer Service Center GmbH
Europaplatz 3, 69115 Heidelberg, Germany

Printed by Libri Plureos GmbH
in Hamburg, Germany